COMMANDER
NICHOLAS

The submarine

The ultimate weapon

DAVIS-POYNTER

First published in Great Britain in 1973 by
Davis-Poynter Limited
20 Garrick Street London WC2E 9BJ

ISBN 0 7067 0079 1 (*cased*)
ISBN 0 7067 0080 5 (*paperback*)

Printed in Great Britain by
Bristol Typesetting Co Ltd, Bristol

Contents

CHAPTER ONE

Under Way on Nuclear Power

On the morning of January 17, 1955, Commander Eugene Wilkinson, USN, stood on the bridge of the newly commissioned submarine *Nautilus*, alongside the Electric Boat Company's yard at Groton, Connecticut and gave orders for the berthing wires to be let go, followed by slow speed ahead. As the vessel gathered way and slid into the main channel of the Thames river, her signal lamp blinked the message 'under way on nuclear power'. It was a historic moment.

The days and weeks that followed were spent in testing equipment, then in May came her first long submerged voyage, from New London, Connecticut to San Juan, Puerto Rico, a distance of 1,381 miles in 90 hours – an average speed of 16 knots and a new endurance record for submarines. She was later to surpass this by a slightly longer passage from Key West, Florida, to New London, at an average underwater speed of 20 knots.

More important than these records was the fact that here, at last, was the true submarine, not tied to the surface like her predecessors by the need to replenish her oxygen supply, but able to remain below indefinitely, or as long as the crew could endure it. This supreme feat, the dream of countless submariners, had been achieved by nuclear power, which could be harnessed to produce energy without combustion and so without the need for oxygen. The *Nautilus* was not merely an improved submarine : she had brought about a change not merely of degree, but of kind. For the first time the submarine could range the oceans of the world and remain hidden.

To appreciate the full significance of this, it is necessary to look more closely at the nature of the submarine as a weapon of war.

The submarine's prime attribute from which all its other virtues flow, is concealment – its ability to hide in the sea. But this advantage of concealment was only gradually achieved : problems lay ahead that were not solely technical. The first battle the submarine had to fight was a moral one. The submarine appeared on the scene before the transition from sail to steam and its adoption as a method of war did not at first commend itself to the world's great sailing ship navies. To those brought up on gun battles at short range, traditional methods of boarding and hand to hand combat, this new weapon with its element of stealth seemed outrageous, if not downright dishonourable : it was hitting a man below the belt : 'unfair, underhand and damned un-English', one First Sea Lord described it.

When the American inventor Robert Fulton, aided by a grant from Napoleon, successfully demonstrated his three-man hand-operated *Nautilus* to the French in 1801 by blowing up an old schooner moored at Brest, the French Ministry of Marine lost its nerve and decided that the crews of such vessels could not be regarded as combatants : they were pirates and, if caught, the British would most likely hang them. When three years later Fulton tried his invention on the British themselves, his ideas were examined by a committee headed by William Pitt, the Prime Minister, whereupon Earl St Vincent, the First Lord of the Admiralty, castigated Pitt for encouraging 'a mode of war which those who command the sea do not want and which if successful would deprive them of it'. This view carried the day and remained the basis of British policy for the next hundred years.

Meanwhile, development of the submarine, particularly in France and America went ahead. By 1900 six navies owned a total of ten submarines with another eleven under construction : France being well in the lead with a total of 14 built and building. This was disquieting news for the Admiralty : St Vincent's policy of discouragement had clearly been overtaken by events and the submarine had become a recognized weapon of war.

The Admiralty now took the standpoint that the submarine was of use only to weaker powers and that the stronger powers need only concentrate on anti-submarine measures. It was Admiral 'Jacky' Fisher who first saw that the submarine was not only the weaker power's weapon, but of value to the stronger power too, and he used all his influence to advance the claims of the submarine and its weapon the torpedo. In a characteristic memorandum he drew a picture of a flock of enemy transports, loaded with troops, with these invisible demons lurking in the vicinity : 'Death near – momentarily – sudden – awful – invisible – unavoidable! Nothing conceivable more demoralizing!'

It soon became apparent that effective countermeasures could not be devised without a fuller knowledge of the submarine itself and the Naval Estimates of 1901 announced the ordering of 'five submarine vessels of the type invented by Mr Holland', to be manufactured under licence by Vickers. The statement added; 'what the future value of these boats may be in naval warfare can only be a matter of conjecture. The experiments with these boats will assist the Admiralty in assessing their true value'.

Once the submarine was fully accepted as a weapon of war, the constant aim of the designers was to improve its powers of concealment, which meant a close attention to hull design as well as methods of propulsion. In shaping the hull, the early designers were better than they knew; John P. Holland's original design, a 'tear drop' form with a single screw, resembled closely the hull form now regarded as the best, evolved after years of experiment at the David Taylor Model basin in Maryland and embodied in 1953 on the USS *Albacore*. It was this hydrodynamically perfect shape, later incorporated in the nuclear powered *Skipjack*, which serves as the model for American and British nuclear-powered submarines today. Holland had been ahead of his time in designing a hull form more worthy of the true submarine than the limited submersible vessel of his day.

The Holland submarine as built for the Royal Navy, was a 105 ton vessel with a four cylinder petrol engine which drove it at 12 knots on the surface, and electric batteries which gave it a submerged speed of 7 knots for 24 miles. But petrol fumes were dangerous and the diesel engine proved safer and more efficient. Britain's first diesel-electric submarine, the *D.1* of 1908, was a larger vessel of 500 tons with twin screws and external water ballast tanks to give it a high buoyancy. A successful type, it set the pattern for the next forty years, including two world wars.

The submarine operations of both world wars show the vessel to have been in reality a surface ship with the ability to carry out submerged attacks, and stay below for a period to evade pursuers. Most of its time was spent on the surface, including the passage to its patrol area; many of its attacks were also carried out on the surface, notably the U-boat 'wolf pack' tactics initiated by Admiral Doenitz against convoys at night. Yet in spite of all the limitations, slow submerged speed and an underwater endurance at the mercy of battery exhaustion, the German U-boats nearly succeeded in cutting the Allied supply lines. The U-boats' defeat was only brought about by intensive counter effort on the part of convoy escorts and aircraft which stripped from the submarine its freedom to operate and replenish on the surface: it was more and more forced to hide, beyond its capacity to do so.

The Dutch invention of the schnorkel or 'snort', as it came to be known in the Royal Navy, gave the submarine a further lease of life. This breathing tube which could be raised by the submerged submarine to draw in air was originally intended for ventilation, but was adapted by the Germans towards the end of the 1939-45 war as a means of operating a submarine's diesels when submerged. This was a considerable advance, and gave the world the 'intermediate' submarine, with a much improved performance; for example, H.M.S. *Andrew* celebrated the Queen's coronation in 1953 by snorting 2,500 miles from Bermuda to Portsmouth.

But this method had its drawbacks. Although not an easy target for radar, the snort tube was larger than a periscope and left a distinct wake; it was also liable to emit smoke and could literally be 'sniffed' by modern detection methods. The submarine was still an air breather, its fundamental nature as a submersible unchanged.

A step nearer the true submarine was taken by the German designer Dr Hellmuth Walter towards the end of the 1939-45 war, using high test peroxide or ingolin which gives up oxygen on being passed over a catalyst. He devised a turbine which ran on a mixture of steam and carbon dioxide formed by the decomposition of hydrogen peroxide; no air was needed. Speeds of over twenty-five knots submerged were possible with this Type XXVI U-boat, but for no more than three hours. The storage of sufficient ingolin proved a difficulty and the Walter boat had to be given diesel-electric power as well – three propulsion systems in all. Although a true submarine only when using ingolin, the type showed promise, and extensive tests were carried out by the Royal Navy with *U-1407* (afterwards renamed H.M.S. *Meteorite*) which fell into British hands at the end of the war. Two experimental submarines *Explorer* and *Excalibur*, with improved high test peroxide engines, were built for the Royal Navy, but further development on these two came to a halt with the spectacular success of the nuclear submarine.

While every credit must be given to American practical

ingenuity for the installation of a nuclear power plant in a submarine, due tribute must be paid to the genius who presided over man's release of atomic energy, the late Lord Rutherford. This large shaggy man, with the look of a modern Viking, propounded in 1911 the nuclear theory of the atom, which was the intellectual guide that showed the way to the release of atomic energy. He encouraged and supported Professor Niels Bohr in his theory of the atomic mechanisms and gave him a place on his staff. One of his pupils was Sir James Chadwick, who later discovered the neutron and became the chief British participant in the invention of the atomic bomb : another was Professor Otto Hahn, who discovered the fission process in uranium, the immediate starting point for the atomic bomb development. Many other eminent scientists who contributed to this were either Rutherford's personal pupils, or under his intellectual influence.

If the introduction of the atomic bomb was the most terrible event of the last war and indeed of the history of mankind, the splitting of the atom also meant energy that could be controlled and harnessed for man's benefit : it could even drive a ship.

Theory began to be translated into action in March 1939, when Dr George Pegram of Columbia University, a leading physicist, proposed to Rear-Admiral Harold G. Bowen, Chief of the Bureau of Steam Engineering, a meeting to discuss practical uses of uranium fission. Also present were Dr Enrico Fermi, a leading expert on the properties of neutrons, Captain Hollis Cooley, head of the Naval Research Laboratory and Dr Ross Gunn, superintendent of the Naval Research Laboratory's Mechanics and Electricity Division.

At this meeting the possibility of initiating a nuclear chain reaction was discussed. A few days later, Captain Cooley and Dr Gunn called on Admiral Bower with a plan for a 'fission chamber' which would generate steam to a turbine for a submarine power plant and appealed for funds for further research. The funds were authorised and that summer Dr Gunn submitted his first report on nuclear propulsion for submarines, pointing out that such a power plant would not require oxygen, 'a tremendous military advantage that would enormously increase the range and military effectiveness of a submarine'.

But war was about to break out, and by the time a feasible method was found of separating the lighter uranium isotope, U-235 – present in mined uranium at the ratio of only 1 to 140 – the atomic bomb had taken priority. Control of fissionable material was almost entirely vested in the 'Manhattan Project' – the code name for The Bomb.

A governmental committee, set up in late 1944 to make recommendations for America's post war policy for atomic energy development urged the authorities 'to initiate and push, as an urgent project research and development studies to provide power from nuclear sources for the propulsion of naval vessels.' A year later, for the first time, the matter got an airing in the Press: the *New York Times* of December 14, 1945 quoted Dr Gunn as saying that 'the basic job of nuclear energy is to turn the world's wheels and run its ships.'

Early in 1946 it was decided to construct an experimental nuclear power pile at Oak Ridge in Tennessee, and the US Navy was invited to take part. One of the four officers selected was Captain Hyman G. Rickover, a former submariner, who had already made a name for himself as head of the wartime electrical section of the Bureau of Ships. A forceful personality with a reputation for ruthless efficiency reminiscent of the Royal Navy's 'Jacky' Fisher, Rickover's name will for ever be associated with the achievement of the nuclear powered submarine and later, the Polaris programme. After reporting to Washington in May 1946 for his new assignment, Rickover lost no time in searching through the files of the

Bureau of Ships for every reference to nuclear propulsion. Among them he found a report by Dr Philip Abelson of the Naval Research Laboratory who planned a submarine on the lines of the fast German Walter type vessel, with his own conception of a nuclear power plant replacing the hydrogen peroxide one. The report stated that 'with a proper programme, only about two years would be required to put into operation an atomic-powered submarine mechanically capable of operating at 26 to 30 knots submerged for many years without surfacing or refuelling'. This high-speed atomic-powered submarine would be able to operate at depths of about 1,000 feet and, 'will serve as an ideal carrier and launcher of rocketed atomic bombs'.

From now on, Rickover devoted all his energies to the idea of the nuclear powered submarine. The naval contingent of four officers became a close-knit team of which he was the head, reporting almost weekly to the Bureau of Ships on various aspects of nuclear power. In spite of powerful backing, the group's final report in the Autumn of 1947 met with a cool reception from the Bureau, who declined to follow its recommendations for a speed up of the naval reactor programme and disbanded the group. This was no more than a temporary setback for Rickover, who remained in Washington at the request of Admiral Mills, Chief of the Bureau of Ships, as his special assistant.

In effect, this move gave Captain Rickover more power than ever, as he was able to acquire a dual personality, one as head of a Naval Reactors branch in the civilian Atomic Energy Commission and one in the Navy's Bureau of Ships, in charge of the Nuclear Power Branch : an advantage he enjoys to this day. He was instrumental in obtaining a signed memorandum from no less a person than Fleet Admiral Nimitz, a former submariner, urging immediate development of a nuclear-powered submarine, which greatly strengthened the Navy's arm, the Atomic

Energy Commission somewhat reluctantly withdrew its objection and the naval reactor study programme was formally approved.

With the growth of Rickover's two-office staff, plans started in earnest. Engineers began to build a land-based nuclear power plant and ship designers worked on a submarine that would house it. Rather than design an entirely new hull *ab initio*, it was decided to take a conventional Guppy class vessel, giving it a more rounded bow to improve underwater performance. It was moreover to be an operational submarine, complete with torpedo tubes, and not purely an experimental vessel.

The world's first power-producing nuclear reactor took shape at Arco, in the Idaho desert, some 600 miles from the sea. The original scheme was to build a large prototype covering some acres of land and later reduce it in size to fit inside a submarine. But the redoubtable Rickover insisted that the land-based prototype must be built inside a submarine hull so that it could be adapted to fit in a sea-going vessel with the minimum of redesign. Rickover also decided to build not one but two plants, Submarine Thermal Reactor 1, the land based prototype at Arco, and Submarine Thermal Reactor 2, the sea-going version which would eventually go into the *Nautilus* : the programme being so organized that each part of the prototype was completed a few months before its counterpart in the sea-going reactor. The submarine hulls for both reactors were built by the Electric Boat Company, of Groton, Connecticut.

In addition Rickover stipulated that a full-sized model must be made so that every component could be seen to fit in its place before starting to build it into either of the operational reactors, although it meant, for a time, countless journeys by the Electric Boat engineers between the model at Groton and the prototype hull, 2,500 miles away at Arco. These moves probably saved the project as much as five years development time.

The Submarine Thermal Reactor was of the pressurized water type. In essence this consisted of a nuclear reactor core contained in a pressure vessel. Circulating water, known as the primary coolant, kept under pressure to prevent boiling, removed the heat generated in the core and transferred it to the secondary, or steam system, where, in a steam generator, it turned normal boiler water into steam to operate the propulsion machinery and generate electric power. In transferring the reactor's heat in the steam generator, the pressurized water was 'cooled' and in turn cooled the reactor by extracting its heat for another cycle.

The reactor had to be encased in heavy lead shielding and similar shielding placed around the circulating coolant water system as this became radioactive in passing through the reactor core.

There had been steam driven submarines before, notably the Royal Navy's 'K' class of 1917, designed to keep up with the speed of the Fleet, but the steam was for use on the surface only giving them a speed of over 23 knots : submerged they used electric drive.

Pressurized water was not the only kind of reactor coolant and heat exchanger capable of providing steam. Parallel with the development of the Submarine Thermal Reactor was a General Electric Company design using liquid sodium and known as the Submarine Intermediate Reactor. According to Rickover, 'at this time, in 1947, we did not know which one would work. As a matter of fact, we thought sodium had a better chance of working than water'. Again there was a two-fold development, SIR 1 being built in a land based submarine hull at West Milton, New York and SIR 2 in a sea-going submarine, later to become the second of the nuclears, the *Seawolf*, slightly larger than the *Nautilus* and completed in 1957. This vessel, while she had many desirable features, also turned out to have inherent maintenance and safety troubles which resulted in severe operational restrictions. Nevertheless, she put up an impressive performance and at one time held the endurance record of 13,761 miles in a submerged run of 60 days.

While design of the Submarine Thermal Reactor was going ahead under the stimulus of Rickover's dual office, a Navy selection board met in Washington to name captains for promotion to flag rank : Among the candidates was Captain Rickover. He was not selected, which meant that should he be 'passed over' the following year he would be automatically retired on completing 31 years of active service.

The contract to build the world's first nuclear submarine was awarded to the Electric Boat Division of General Dynamics Corporation on August 21, 1951 and the following December the Navy announced that it would be named *Nautilus*. This was a name that breathed innovation. Even those unaware of Fulton's early venture, were familiar with the name from Jules Verne's phophetic tale *Twenty Thousand Leagues Under The Sea*, written in 1869. President Harry S. Truman himself laid the keel at the Groton shipyard on June 14, 1952, his initials being welded into the keel plate. At the ceremonies that followed the Chairman of the Atomic Energy Commission made special mention of Captain Rickover, 'whose energy, drive and technical competence have played such a large part in making this project possible'. The following month, Rickover was awarded a Gold Star in lieu of his second Legion of Merit – he had received the first for war service – the citation paying tribute to his tenacity against frustration and opposition. 'He has constantly advanced the submarine thermal reactor well beyond all expectations : his efforts have led to the laying of the keel of the world's first nuclear-powered ship well in advance of its original schedule.'

The day after the presentation of the award, the Navy selection board met again to consider Rickover, together with several other engineering specialists, for promotion to Rear Admiral. He was passed over for the second and last time. Such a storm was raised in the Press and among Government officials that the Secretary of the Navy was forced to take action. He gave orders that Captain Rickover was to be retained on the active list for another year, at the end of which the selection board was to meet again to promote to Rear Admiral one engineering specialist captain experienced and qualified in the field of nuclear propulsion. Rickover got his promotion.

The subsequent career of this remarkable man, who has been called the 'father of the atomic submarine', has been no less unusual. Promoted to Vice-Admiral in October 1958, he was retained on the active list, on President Kennedy's orders, until February 1964 – two years beyond his normal retirement date. It was then decided to keep him on duty more or less indefinitely. His dual position, his energy and determination in carrying through the nuclear submarine programme, his prestige with Congress and the press have given him unparalleled authority not only in the design of naval nuclear reactors but also in the selection and training of officers for nuclear submarine duty.

This century has given us three great Admirals, who in a real sense were 'shapers of navies'. Unlike the fighting Admirals – such as Nelson, Jellicoe, Beatty or Cunningham, who became famous in battle, the main achievements of the 'shapers' were in peacetime. The three who will go down in history as the great naval innovators, fashioning entirely new navies, are Britain's Fisher, Soviet Russia's Gorschkov – of whom more will be said later – and America's Rickover.

A red-letter day in the history of the nuclear submarine was March 30, 1953 when the land-based submarine thermal reactor at Arco, Idaho 'went critical', that is, a self-sustained controlled nuclear chain reaction took place within the reactor. Six weeks later the rest of the propulsion system was connected up and power was very gradually increased until June 25 when full power was reached. Most engineers would have been satisfied with a 24-hour or, possibly a 48-hour run at full power. Not so Rickover, who had set his heart on a simulated Atlantic crossing, which would capture the imagination of the world. The naval 'crew' began regular four hour watches and a 'course' from Nova Scotia to Ireland was plotted on the chart.

The two day mark was passed without incident : never before had any submarine 'travelled' so far at full power. At sixty hours troubles began, first carbon dust in the generators, then one of the vital water pumps gave cause for anxiety, followed by failure of a tube in the main steam condenser. Not surprisingly, the Westinghouse representative recommended an immediate halt. But Rickover held on; twice the plant had to be throttled back to half power and once to two-thirds power, but there was no stoppage throughout the 86 hour run. The gamble had paid off : the land-locked vessel had 'travelled' 2,500 miles at unprecedented speed. Moreover, when the reactor cooled down after the run and could be inspected, it was found to be free from any serious defect. The *Nautilus* had not yet been launched and its first sea voyage was more than a year away, but the land-based submarine at Arco had given undoubted promise of things to come.

The *Nautilus* was launched on January 21, 1954 after one year, seven months and seven days on the building ways, Mrs Eisenhower performing the naming ceremony. Apart from her propulsion system she had cost $29,000,000 (£12 million). Her six 21 inch torpedo tubes made her an effective fighting ship, but with a degree of comfort for her crew exceeding that of any previous undersea craft.

Most important of all was the question of pure air, which the *Nautilus* initially obtained by a combination of bottled oxygen and a device known as a 'CO_2 scrubber' which removes harmful gases from inside the submarine. As long as submarines had to surface at frequent intervals for the re-charging of batteries, or were drawing in air through a 'snort', this was no great problem. But the extended patrols now possible with the nuclear submarine gave an impetus to improved methods of revitalising the atmosphere. Chief among these was the maintenance of the submarine's own oxygen supply by electrolysis of sea water. Equipment for this is now standard on all nuclear submarines and makes bottled oxygen unnecessary.

Fumes from tobacco and cooking are dealt with indirectly by the 'CO_2 scrubber', for these and other carbon monoxide gases are passed through a special burner which turns them into carbon dioxide, which the 'scrubber' can then attack.

Thermometers in nuclear submarines are of the alcohol type to obviate the harmful 'trace gas' from mercury vapour. Care too has to be taken in the design of refrigerators and in the use of aerosol sprays, to minimize the danger from Freon vapour. Oil base paints and certain cooking fats, are banned.

Danger to the 95-man crew from the close proximity of the radioactive uranium that served as the *Nautilus*'s fuel was obviated by heavy shielding. In fact, a nuclear submarine's crew receives less radiation than they would normally from natural sources ashore.

The *Nautilus* was commissioned as a United States ship on September 30, 1954. Her nuclear propulsion plant began basin trials two months later, developing full power on January 3, 1955. This was an almost exact fulfilment of the plan drawn up by the Navy five years before, that the submarine should be ready for sea on January 1, 1955. A fortnight after achieving full power alongside she left harbour for the first time under nuclear power.

As the first true submarine, the *Nautilus* had achieved in one stroke not only the perfect concealment necessary to fulfil its role, but an entirely new form of mobility, not to be found on land : No land vehicle has yet been given a nuclear role. Mobility on land, which has furnished such a large element of surprise in the past, is becoming increasingly difficult, as methods of surveillance such as radar, high altitude reconnaissance and satellite observation are improved. Dependent for the most part on petrol or diesel driven transport, troop and artillery movements call for a vast build up of fuel supplies before any major attack. Free of all this, nuclear power's mobility would seem to offer unusual advantages to surface vessel and submarine alike.

In a world of dwindling overseas bases, a strong case could be made for nuclear surface warships and the Americans were not slow to exploit it. The USS *Long Beach*, 14,200 tons, laid down in December 1957 was the first surface ship in the fleet to have nuclear reactors as her power plant, two of these gave her a speed of over 30 knots. She cost $332,500,000 – the most expensive cruiser ever built. She was followed by the USS *Enterprise*, 75,700 tons, which cost over twice as much as the conventional *Forrestal* class, but for this extra cost carried 25 per cent more warlike stores and twice the aviation spirit of the earlier ship and was able to operate up to 100 aircraft.

Then came the nuclear-powered frigate *Bainbridge*, 7,600 tons, the forerunner of the *Truxton* and later classes. A second nuclear powered carrier, the *Nimitz*, due to enter service in 1972, will have only two reactors, but developing the same power as the *Enterprise*'s eight : an indication of the improvements in reactor design. In 1964, the *Enterprise, Long Beach* and *Bainbridge* formed the world's first all nuclear task force, completing a round-the-world cruise of 30,500 miles in just over two months.

The high initial costs deter most Governments from embarking on a programme for a nuclear surface fleet, but there is no gainsaying its advantages. Admiral Rickover has estimated that the Seventh Fleet could be reduced to one third its size and still retain the same striking force by providing the warships with nuclear power and eliminating most of the fleet auxiliaries and support vessels. Long range costing has shown that taking all the logistics into consideration over a normal twenty year life span, the nuclear warship costs no more than its conventional counterpart. No clear case has yet been made for large scale adoption of nuclear propulsion in merchant ships since these vessels of their very nature must make port frequently.

All the obvious advantages of nuclear power are multiplied in the case of the submarine where nuclear propulsion enables it to fulfil its true function. Logistically free, and no longer obliged to observe an 'economical speed' dictated by Government economy or by its own need to conserve fuel, it can traverse the oceans at consistently high speeds in the calm depths below the turmoil of wind and weather. This capability of high speed in all weathers gives the nuclear submarine the edge over the surface vessels seeking to destroy it; the submarine can lose itself in the 'ever changing sea'. With the *Nautilus*, the nuclear submarine and the sea seemed made for each other, in a manner almost surpassing the wonders of space travel. Man landed on the moon without the aid of nuclear power. But the sea was to be the one place where nuclear power would revolutionize everything.

CHAPTER TWO

The Polaris Submarine

Less than a year after the *Nautilus* went to sea, developments began which were to bring together the nuclear submarine and the ballistic missile, resulting in the most powerful weapon system ever devised. The Office of Naval Research in early 1955 commissioned the General Dynamics Corporation to examine the possibilities of a submarine missile system. The result later that year was a comprehensive report: 'Strike Submarine Missile Weapons Systems Study', which proved, at least on paper, the possibility of a complete submarine strategic offensive force. With remarkable foresight, it stressed the advantages of a solid propellent for the missile and of a ship-borne inertial navigation system for the submarine.' As a result of this, in November 1955 the United States Defence Department ordered the Army and Navy to collaborate in the development of a 1,500 mile ballistic missile. Soviet Russia had exploded her first H-bomb and the warning was clear; the whole strategic situation had changed: the two most powerful nations were becoming capable of destroying each other with inter-continental ballistic missiles against which there was no known defence. The enemy might thus be tempted to win by a so-called 'pre-emptive' attack. It was vital to find a 'second-strike' weapon that would survive the critical onslaught and be available to retaliate. What better than a force of missile-carrying submarines, which could inflict unacceptable damage on an enemy even after land and air forces had been destroyed? A single undetected submarine could deliver many megatons of high explosive in the thermo-nuclear warheads of its rockets. This was the nuclear deterrent par excellence for with it there would be little temptation to risk starting a war in the hope of delivering such devastating damage in the initial attack as to prevent any retaliation. The matter had now become a national priority.

The Americans had already fired some captured German V2 missiles at sea, but not from a submarine. The 12-ton, 46-foot long rockets which had cost 2,754 British lives and injured 6,523 more, before Allied invasion forces overran the launching sites, were most unsuitable for submarine use on account of their size, the dangerous nature of their fuel – a mixture of liquid oxygen and alcohol – and the difficulties of storing it at a sufficiently low temperature.

The first outcome of the Army-Navy collaboration was the Jupiter missile, 60 feet long and liquid fuelled. It was too large and dangerous to handle for the Navy's liking, but in default of a sufficiently powerful fuel propellant the Navy determined to make the best use of it and adapt the Jupiter for naval use. A new department was formed, known as the Special Projects Office – later to become the Strategic Systems Project Office – with direct access to the Secretary of the Navy, and responsible for the submarines built around the weapon, their manning and all support facilities.

Its first director was Rear-Admiral William F. Raborn Jr, a naval aviator who had served with distinction in the Pacific during the war, and was an expert in management techniques. Under him as Technical Director was Captain Levering Smith, an engineer officer and ordnance expert who as a Commander some years before had worked on a solid fuel rocket at the Naval Ordnance Test Station at Inyokern, California. Levering Smith was to prove the brains behind the Polaris weapon and its subsequent conversion to Poseidon. Now a retired Rear-Admiral, he heads the Strategic Systems Project Office;

means having been found to retain his services in the Rickover manner.

Development work on the Army's Jupiter missile was carried out at Huntsville, Alabama, and a naval team from Special Projects was soon installed to design a seagoing version. Among other scientists they found there were German missile experts led by Wernher von Braun. The Navy wanted the missile as short as possible for easier ship handling and eventually its length was cut from 65 feet to 60 feet, but to compensate for this its diameter was increased from 95 to 105 inches.

Plans were drawn up for the smallest possible submarine that could handle the missile – a vessel of about 8,000 tons, nearly three times the size of *Nautilus*. Four Jupiters would be carried, housed in tubes that extended through the conning tower, and to fire them, the submarine would have to break surface. A solid fuel for the missile had the serious drawback of too low a thrust, which could only be overcome by using a cluster of rockets, with the disadvantage of increasing the size of the missile still more.

But Captain Levering Smith's Special Projects team persevered with solid fuel experiments and in mid 1956 started work on a small but promising solid fuel missile. Such a missile, coupled with one of the smaller warheads then being designed offered untold advantages to a submarine : it might even be fired when submerged. A decision now had to be made as to whether to concentrate on the solid fuel missile or carry on with the Jupiter project. Admiral Raborn decided to go for the solid fuel missile and quickly got the Navy Department's approval. It took rather longer to get the Defence Secretary, Charles E. Wilson to agree, but eventually the Navy ended its participation in the Jupiter programme and on January 1, 1957 the Polaris ballistic missile programme was formally approved.

Lockheed Aircraft were given the main contract for the

Polaris missile and a group of scientists and engineers from the Massachusetts Institute of Technology worked on the missile navigation system. These and other industrial firms were brought together by Captain Levering Smith to formulate plans for a missile system that would be ready by 1963. The result : Polaris would be only 28 feet long, weigh 28,000 pounds and have a range of 1,500 miles. This at once put the nuclear submarine in business as a ballistic missile carrier since a vessel twice the size of *Nautilus* could carry 16 Polaris missiles, leaving room for crew space, fire control, navigation and other standard equipment.

Among the many problems facing the Special Projects team two in particular may be mentioned. A means had to be found to launch the missile from underwater to a position clear of the surface when the rocket motor could be ignited – it would obviously be too dangerous to start the rocket inside the submarine. The answer they arrived at was compressed air : the missile would start its journey like a vertical firing torpedo. Later, high pressure gas was to be used – and eventually steam.

Extremely accurate navigation without the possibility of an optical 'fix' through a periscope or electronic aids via an aerial had also to be devised. Even a small miscalculation in plotting the submarine's position would put the missile off target : an error of a degree in direction would mean a miss of many miles at the other end. The answer was found in the Ship Inertial Navigation System (SINS) which uses the gyroscope's property of holding to the same position irrespective of the submarine's movement. Using this as a reference point, every change of course, depth and speed is fed into a computer which gives an exact navigational position at any time.

A miniature inertial guidance system was also fitted to each Polaris missile, which would receive an 'instruction' immediately before firing. The missile itself was to be a two-stage one, that is, two separate rocket motors to supply the necessary thrust. After the first few seconds of flight the burned-out lower motor would drop off, followed a little later by the burned-out second stage.

A further stimulus to the Polaris programme was given by Russia's space rocket achievements : two satellites had been put into orbit round the earth before the end of 1957, clearly indicating that their rockets were capable of delivering the Bomb on almost any American city. It became vital to advance the date of the first Polaris submarine becoming operational, which the Special Projects office announced could be done – by as much as three years – with certain reservations. Instead of a 1,500 mile range, the first Polaris to go to sea would only rate 1,200 nautical miles. Moreover, the first Polaris submarine would now have to be ready in two years instead of five. Many doubted whether this was at all possible : the latest nuclear submarine, the *Skate*, had taken 29 months, and a work-up period of some months on top of that was needed before she could become operational. A quicker method would be to take a submarine already on the stocks, cut her in two and add a mid-section containing the missile tubes, fire control and navigational equipment. A suitable vessel was at hand in the *Scorpion*, a 252 foot attack or hunter-killer submarine building at the Electric Boat Company's Groton yard. To give her the additions called for by Polaris, would require an extra 130 feet, shift work was at once started in the *Scorpion*, and on a second un-named sister vessel, all round the clock and nearly seven days a week. Further orders brought the number in the class up to five.

The 'second generation' Polaris submarine designed as such from the start, the *Ethan Allen*, was a 410 foot vessel with extra crew accommodation and space for improved SINS and fire control equipment, ordered from the Electric Boat Company. Meanwhile, test firings of the first Polaris missile took place in the autumn of 1955, but it was not till the following April that a fully successful shot was fired.

The first Polaris submarine, the re-constituted *Scorpion*, now renamed the *George Washington*, was launched on June 9, 1959. Speaking at the launching ceremony, Wildred J. McNeil, an Assistant Secretary of Defence, said that the vessel, with its Polaris missile, 'is the first naval weapon system to be specifically designed for strategic employment against land targets and adds an entirely new dimension to our naval power.' Of its destructive power, he said; 'no attacker could hope to escape retribution, even given the advantage of striking the first blow'.

On December 30, 1959 the *George Washington* was commissioned. She successfully carried out the first launch of a Polaris test missile from a submerged submarine off Cape Kennedy, Florida on July 20, 1960 and the message was sent to President Eisenhower 'Polaris – from out of the deep to target. Perfect.'

Two days later, the Navy awarded contracts for three more Polaris vessels, the *Thomas Jefferson*, fifth and last of the *Ethan Allen* design, the *Lafayette* and the *Alexander Hamilton*, which started the third generation of Polaris submarines. Thirty-one *Lafayettes* were eventually to be ordered : vessels with better crew facilities and an increased overall length of 425 feet. Steam power replaced the more bulky compressed air system for launching the missiles in later vessels of the class and improvements in fire control considerably reduced the time needed to fire all sixteen missiles, which in the earlier classes must have been fifteen minutes.

The first Polaris deterrent patrol began when the *George Washington* put to sea from Charleston on November 15, 1960. This was done with much publicity and a Navy band playing a stirring Sousa march, on the score that to be effective a deterrent must at least be made known. Later patrols were to be started in secret.

The *George Washington*'s first patrol lasted sixty-six days: and was described by her captain, Commander James Osborn as a success, her missiles he said had been 'cocked for important use' throughout. The year 1960 ended with a total of two combat-ready Polaris submarines at sea, for the *George Washington* had been joined by the *Patrick Henry*.

John F. Kennedy, who became President of the United States in January 1961, showed himself to be remarkably Polaris minded. He had not been in office a fortnight before he had authorized the immediate construction of five Polaris submarines: shortly afterwards he asked Congress for ten more in the fiscal year beginning July 1, 1961 – making a total of 29. He also ordered a speed-up in the development of the missile itself. The A-1 fitted in the *George Washington* class had a range of 1,200 nautical miles, the A-2 then under development, had a range of 1,500 miles, while the projected A-3 would have a range of 2,500 nautical miles. President Kennedy's accelerated programme aimed at getting A-2 to sea in late 1962 and A-3 in 1964. Conversion of the earlier submarines to A-2 was to be carried out in the dockyards.

The US Navy, assessing the 'combat readiness' of the missiles in these Polaris submarines revealed that while on patrol, with a 15 minute warning period, all 16 missiles were ready to fire 95 per cent of the time, while 15 of them were available 99.9 per cent of the time; the very slight drop in readiness being due to inspections and tests of the missiles or electronic equipment.

Early in 1961 the US Navy with the agreement of the British Government decided on an advanced European base for their first Polaris submarine squadron and sent the USS *Proteus* to Holy Loch on the Clyde. She carried a large crane for handling missiles, reactor cores and nuclear waste containers and among other support facilities, a large technical library. A shuttle service of specially equipped cargo ships would bring missiles and other supplies to the *Proteus*. A few days after her arrival the *Proteus* was able to minister to the *Patrick Henry*, America's second Polaris submarine, which returned from a submerged patrol of 66 days and 22 hours, bettering by half a day the *George Washington*'s first cruise record.

The *Proteus*, a supply vessel specially converted for the support of nuclear submarines, was relieved at Holy Loch two years later by the submarine tender *Hunley*, designed and built for the purpose. In the interests of making the deterrent world-wide, support facilities were also arranged at the Pacific bases of Pearl Harbor and Guam.

It was now time for Britain to consider her own position with regard to the deterrent. Apart from the submarine-launched ballistic missile and its land-based counterpart, the deterrent could be wielded by the manned bomber, and plans were made for joint Anglo-American use of the Skybolt air-launched weapon, which could be launched 100 miles from the target. But the project, never very much favoured by the Kennedy administration, did not materialize and after a series of test launches had failed, Skybolt was cancelled. This left Britain without a strategic deterrent of her own, a matter for urgent talks between the two countries.

In late December 1962 President Kennedy and the British Prime Minister, Mr Harold Macmillan, met at Nassau in the Bahamas. They agreed, that with the concellation of Skybolt, the United States would provide Britain with the Polaris weapon system and that Britain would design and build the submarines to carry it and provide the nuclear warheads for the missiles. It augured well for the future that attending the conference on the British side was Vice-Admiral Michael Le Fanu, Third Sea Lord and Controller of the Navy, whose dynamic support ensured the foundation for the British Polaris programme, to be described later by the Navy Minister as 'the toughest peacetime task, in a given time scale, which the Navy has ever been handed'. It was not the least of Admiral Le Fanu's achievements.

A month after the signing of the Nassau Agreement, the British Government decided on a force of four or five nuclear powered submarines, each equipped with 16 missiles, the first to be operational by June 1968 and the remainder to follow at six monthly intervals, with full support from a shore base available by mid 1970.

A Polaris Executive was formed under Vice-Admiral Hugh Mackenzie to co-ordinate the whole programme, which included the design and production of the missiles' warheads. The submarines themselves were designed at Bath by a team led by Sir Rowland Baker, one of the Royal Corps of Naval Constructors' most brilliant operators. The order for four vessels was placed in May 1963 and the *intention* to build a fifth confirmed in February 1964.

The Royal Navy is no stranger, even in peacetime, to 'crash' building programmes. In October 1906 the battleship *Dreadnought* had put to sea a year and a day after keel-laying, ushering in a new era, that of the all-big gun, turbine-driven warship. But it was thanks to the relentless drive of Admiral 'Jacky' Fisher that this lead had been secured for Britain. Now with Polaris the Royal Navy was to take another decisive step, calling for even greater dedication.

The plan to build five Polaris submarines was dropped in February 1965 by the Wilson Government for economy reasons and to make it possible to increase the number of nuclear fleet submarines. Five would have guaranteed two submarines on patrol at any time, now the Navy had to be content with one and a half. At a time of pressing demands for additions to the conventional fleet, guided missile destroyers, frigates, assault ships, helicopter-carriers and so on, much crystal-gazing would have been needed to prove this balance an uneven one. Even the ultimate deterrent cannot be the only weapon in a nation's armoury.

A late start with Polaris had one great advantage : the programme could be planned well in advance. The Americans, with characteristic vigour and despatch, had completed theirs two years ahead of schedule. But they had to improvise as they went, grafting new methods on to conventional submarine building. At Vickers and Cammell Lairds, both chosen to build two of the British vessels, it was possible to select the most suitable tools and equipment from the start. This involved, particularly at Cammell Lairds, new shops and drawing offices, welding bays, overhead travelling cranes, giant mandrels and special transporters to move the hull sections from welding bay to slipway – much of it in advance of anything the Americans had got. Much of the work, for instance, could now be done under cover in the welding bays. When the *Renown* was laid down at Cammell Lairds in June 1964, a prefabricated section weighing 160 tons was moved onto the slipway – some three times larger than the section used for the *Dreadnought*. Soon sections of 240 tons were to be moved from welding bay to building slip.

In one respect American experience was invaluable – the overall control of the programme. Programme Evaluation Research Task and Review Technique (PERT) was the means whereby the latest computer techniques were harnessed to assist, not only in the construction of the submarines themselves, but in the setting up of all the ancillary services including the new submarine base. PERT gave a graphical display of all the actions needed to achieve any particular aim or milestone, such as, delivery of equipment to the shipyard, construction of the submarine itself, fitting of the nuclear power plant, the sea trials and testing of the weapon system; all these events were examined from the viewpoint of the longest time to completion. Once this was known, additional effort could be directed to the part of the programme that would best hasten progress. PERT was the greatest aid to planning in the technical field that had ever been produced. But with the hull, propulsion system and most of the equipment, bar the weapon system, built in Britain, the vessels themselves were truly British. It was not true, as many people believed, that a whole American midship section had somehow been inserted between a Dreadnought-type bow and stern.

The building of nuclear and Polaris submarines had brought about a revolution in British shipbuilding methods. The absolute dependability demanded of the Polaris submarine on patrol dictated a hitherto undreamed of stress on quality. It was no longer enough to enjoin the shipbuilder to supply materials 'without defect', 'to best practices' or 'to the satisfaction of the customer'. In a Polaris contract the shipbuilder himself was required to set up a quality control organization to assure the Chief Polaris Executive of the quality of his own and his sub-contractors' products. All this was in addition to the work of the independent monitors or naval overseers, appointed for duty with the two firms.

Nuclear submarines have a high underwater speed and a well-nigh perfect hydrodynamic form. They can bank, rise and dive like an aircraft. Unlike the aircraft, however, the submarine has to operate within the comparatively narrow limits set by the diving depth. Clearly a failure of hydroplane control at speed could have the direst consequences; hence the imperative demand that such control systems and their emergency alternatives would be built and installed to a rigorous specification. Again, the value of the Polaris submarine would be nil if the weapon system was unserviceable when required. However perfectly manufactured, a complex modern weapon system relies heavily on such ship systems as electric power, hydraulics, fresh and chilled water, high-pressure air, ventilation and air-conditioning. The weapon system is only as strong as the ship and the systems which support it. In short, everything possible had to be done to prevent the abandonment of a patrol due to any type of failure. The task was exacting, not only for the designers, but for the 800 odd commercial firms working to meet their demands.

Building the submarines was only part of the challenge. The original estimate of £353 million was for a force of four vessels and all its support spread over a nine-year period. This implied a base, workshops, a Polaris training school, floating dock, armament depot, married quarters, recreational and welfare facilities, all to be completed to a rigorous timescale. With the American advanced base at Holy Loch, the Clyde was already associated with the Polaris project. Easy access to the Atlantic, deep-water facilities, wharfage, the proximity of existing naval establishments, housing space for families, all indicated the Clyde area.

On the northern side of the Clyde, the Gareloch, seven miles long and just under a mile wide, has long been an important deep-water anchorage. Its naval importance goes back to the days of Robert the Bruce and King James IV, both of whom used it as a naval base. Faslane Bay, in the northern half of the loch is sheltered behind the natural breakwater of the Rhu narrows and is deep enough for the largest of ships. It gained importance during the Second World War when it became the emergency port for Glasgow. Its six 500 feet berths, large marshalling yards, warehouses and cranes helped to mount the North African campaign. Faslane thus became the natural choice for a submarine base, and for some years before it had been the base of the Third Submarine Squadron, consisting of *Porpoise* and *Oberon* class vessels, with the later additions of the nuclear fleet submarines *Dreadnought* and *Valiant*, and a depot ship, the *Maidstone*.

It has been argued that the choice of such a busy focal area for shipping carried grave security risks, in that departures of submarines for patrol could be observed and the vessels tracked. The reverse is true. A submarine can leave undetected at night, while the busy area is a help rather than a hindrance. What little noise the submarine may make in those conditions is soon merged with the incessant underwater pulsation and chatter of coastal shipping. There is no better 'get-away' than to get lost in the crowd.

The first unit of Britain's Polaris programme to be completed was the £9½ million Polaris school which dominated the base like a medieval fortress and keep. Until then the training of the first weapon crews and of the staff for the school had been carried out at the US Navy's guided missile school in Virginia, where the result, for the British officers and ratings was well above the American school's average. The school contains a full scale missile tube, complete with an inert missile, and 15 other tubes, with the same equipment as that fitted in the missile control and navigation centres in a submarine. Every situation that could conceivably arise on board, as well as some that could not, was to be rehearsed at the training school.

Of equal if not greater importance than the construction of the submarines and their base were the preparations for the manning and well-being of the crews. To obtain the maximum availability of the vessels, two crews were necessary : for the Americans Blue and Gold, for the British Port and Starboard. When a Polaris submarine returns from her two month patrol, there are three days of turnover, after which the new crew takes over the vessel completely and prepares her for the next patrol. The old crew after 14 days' leave, come back to the school for refresher and advanced courses, after which they go to the workshops to back up the base weapon and repair staff. An important factor was the provision of homes and housing estates nearby. These have been built to accommodate not only the families of officers and men of the submarines but also the civilian workers who help to run the Faslane base. Even when on patrol a man is not entirely cut off from all news from his family. Submarines can receive messages when submerged, and a system of 'familygrams' ensure that the next of kin of the 144 man crew can send messages of up to 20 words a week per man. But the womenfolk clearly have the upper hand – the men cannot answer back!

Britain's first Polaris submarine, the *Resolution*, put to sea on her first patrol in June 1968, matching the target date set five and a half years earlier. In itself this was a near miracle, considering the complexity of the programme, the new skills demanded of the shipbuilder and the 800 sub-contractors – to say nothing of the odd strike. This underwater giant – as large as a County class guided missile destroyer, loaded with her 16 ballistic missiles with a range of close on 3,000 miles, slipped her moorings and left unobtrusively for an unknown destination in the ocean, there to remain *incommunicado* for two whole months, for this was an operational patrol and it is normal for a submarine to preserve radio silence.

This well-nigh monastic separation from the world was well described 200 years ago by Captain James Cook, setting out for his second voyage of discovery in an earlier *Resolution*: 'I can hardly think myself in this world so long as I am deprived from having any connections with the civilized part of it, and this will soon be my case for two years at least.' Two years above the surface, or two months below it: who is to say which is the harder task?

The *Resolution*'s departure did not mean the immediate handing over to the Royal Navy, in its entirety, of the nation's contribution to the nuclear strategic forces of the West: this was to be a responsibility shared with the Royal Air Force for another 18 months, by which time all four vessels of the Polaris force would be ready.

Repulse, *Renown*, *Revenge* were still to come famous names, long associated with capital ships of the past. This was logical and befitted their size and power, for these submarines were as different from their predecessors of the last war as the P & O cruise liner from the coastal tramp.

The wielding of the ultimate deterrent was certainly a new role for the Navy : but only in the sense of a new form of the Navy's traditional peacekeeping function. The *Resolution* on her two months' vigil in the depths of the ocean was performing a service not very different from that of Nelson's weather-beaten ships keeping watch off Toulon. Nelson's ships were there to prevent the French and Spaniards joining forces to the detriment of European peace. *Resolution* was on patrol to prevent the unleashing of catastrophic destruction on the free world.

It has been estimated that over the past 3,000 years major armed conflict has broken out on an average every 34 years. There have, of course, been a few larger intervals of peace – seldom more than 100 years – but by and large, the general pattern remains the same. The 34 year cycle in the confrontation of strong versus weak nations has a pattern that is almost invariable. Starting with threats, there follows panic, arming, war, then disarming and extreme weakness. After this, more threats, panic, arming and so on, all over again. This teaches the lesson that deterrence must be kept at a constant and sufficient level to keep the aggressor at bay. The bully will never stand up to a man his own size, or one able to hurt him. While it may be difficult to determine what is sufficient, at least the deterrent must be seen to be there.

On the occasion of Britain's first Polaris submarine leaving for her first patrol, *The Daily Telegraph* said in a leading article 'The *Resolution*, in making the first dive of her patrol into the waters of a troubled world, will be taking out on behalf of the nation, the best insurance policy it has ever had.'

CHAPTER THREE

The Capital Ship: The Nuclear Fleet Submarine

The Polaris submarine was a vessel apart, a strategic warfare ship. If its weapon, the long range ballistic missile, ever had to be used, the submarine would have failed in its purpose. True, it had a secondary armament of torpedoes, but these were for self-defence, or for use against enemy submarines, or in some exceptional situation demanding the vessel's use as an ordinary submarine : all outside its normal function of deterring the nuclear onslaught.

Sea warfare was to receive an even more spectacular – if less catastrophic – impact from another version of the true submarine, known as the attack, the hunter-killer, or the fleet submarine – all describing various aspects of the nuclear-powered, conventionally-armed vessel. Here was an underwater menace as fast or faster than any surface ship, of unlimited endurance, deep diving and far ranging, able not only to do supremely well what ordinary, air-breathing submarines were already doing, but to keep on doing it. Armed with anti-ship missiles as well as torpedoes this type of submarine was soon referred to in naval circles as the 'capital ship' of the future, a term hitherto reserved for battleships or aircraft carriers.

The US submarines *Nautilus* and *Seawolf* had taken nuclear power to sea, the first with a pressurized water plant, the second with a liquid sodium one – later to be replaced by the spare set of pressurized water machinery earmarked for the *Nautilus*. Both had convincingly shown that nuclear power could provide the motive power for a true submarine. Experimental vessels in the sense that for a period they had been test beds for differing types of nuclear propulsion, they had also been built with torpedo tubes and sonar and later became fully operational units of the US Fleet. In this they differed from Britain's high test peroxide submarines, *Meteorite*, *Explorer* and *Excalibur* which had been built purely for experimental purposes.

The way was now clear for a production line of nuclear-powered attack submarines and four were laid down by the US Navy all of which were completed between 1957-59. Similar in design to the *Nautilus* and *Seawolf*, the *Skate* class were shorter by over 50 feet and displaced 2,550 tons, nearly 1,000 tons less than *Nautilus*. Submariners have traditionally tried to keep their 'boats' small, more manoeuvrable, and offering smaller targets for detection, and this class faithfully reflected the trend. But although shorter than many conventional submarines, their displacement was about half as great again, due to the larger hull diameter required to house the nuclear power plant.

They were given improved and somewhat smaller pressurized water reactors than the *Nautilus*, with easier access for maintenance. The *Skate*'s shakedown cruise in 1957 from New London, Connecticut to the Royal Navy's base at Portland, Dorset in eight days eleven hours set up a new west-to-east submerged speed record. Her return voyage from the Lizard to Block Island was even better, seven days five hours. Thoughts now turned to a convincing demonstration of the versatility of the attack submarine, which could also have profound implications for future warfare – a voyage under the Arctic ice.

In any future conflict between Russia and the Western Alliance, the Arctic Ocean would be of supreme importance. With an area five times as large as the Mediterranean in which powerful nuclear-propelled vessels could deploy, the Arctic might play a similar role to that of the Pacific in the war with Japan. The true submarine had widened the strategic importance of the oceans to include areas hitherto barred by ice.

The first attempt to reach the North Pole by submarine had been made by the late Sir Hubert Wilkins in 1931, using a former American submarine O-12, which he renamed *Nautilus*. Starting from Bergen, he made for the ice pack between Spitzbergen and Greenland, but beset by mechanical troubles had to return to base after voyaging further north than any previous vessel under its own power.

During the war British submarines and German U-boats had also experimented with limited under-ice navigation. In 1947 the American submarine *Boarfish* penetrated six miles under the Arctic ice pack, with Dr Waldo Lyon of the US Naval Electronics Laboratory on board, who later devised an upward-looking echo sounder for use under the ice. In 1952 the US submarine *Redfish* went over 20 miles under the ice pack, remaining submerged under the ice for nine hours.

Shortly after Commander William Anderson had relieved Commander Wilkinson as Captain of the nuclear-powered *Nautilus* in June 1957, plans were made for the vessel to make an 'ice probe', as far as the North Pole on her way to the Eastern Atlantic for a large scale NATO exercise. A trial run to check equipment and get the feel of operating under the ice met with unexpected trouble. Almost at her turning point 50 miles under the ice and while surfacing in an opening, the *Nautilus* severely damaged both periscopes against an undetected floating block of ice. One periscope was beyond repair, the other developed a bad crack. After emergency repairs a second attempt was made, this time foiled by gyro trouble : but the *Nautilus* had been under the ice for 74 hours and covered nearly 1,000 miles. A third attempt had to be abandoned for lack of time before the NATO exercise.

Plans were then made for a Polar voyage the following year, in which the *Nautilus* would sail from the Pacific to the Atlantic via the frozen Arctic Ocean. All preparations were carried out in secret as the vessel would be operating in an area where there might be Soviet submarines. An armoured cap was given to the top of the conning-tower or sail, to protect the heads of periscopes and masts from ice – a precaution that was later extended to all attack submarines – and an inertial navigation system and extra sonar gear installed.

On June 17, the *Nautilus* passed through the Bering Strait, which separates Alaska from Soviet Siberia. Soon after entering the Arctic Circle she submerged and slowly crept along beneath the pack ice, in one place passing a few feet beneath an inverted peak projecting 63 feet below the surface. Even greater projections lay ahead, leaving insufficient room between the *Nautilus*'s keel and the ocean bed and once more the attempt had to be abandoned, the *Nautilus* returning to Pearl Harbor. After a preliminary area reconnaissance, the *Nautilus* left Pearl Harbor again on July 23 and this time met better conditions, reaching the North Pole on August 9, the first vessel in history to do so, a feat duly recorded by Commander Anderson in the message: '*Nautilus* ninety north'. The submarine's captain was awarded the Legion of Merit and the vessel herself the first Presidential Unit Citation ever awarded in peacetime. On her way back to New York the *Nautilus* set up another submarine record for the Atlantic crossing: just under six days 12 hours, an average speed of 21 knots. A tremendous welcome awaited her at New York and during the celebrations Admiral Rickover was presented with a block of Polar ice.

The *Skate* followed closely on *Nautilus*'s heels, with orders not to go under the ice until the *Nautilus* came out. Commander Calvert set course at the end of July for the passage between Spitzbergen and Iceland. As the

Skate neared the ice pack her crew learned for the first time over the radio that *Nautilus* had reached the North Pole a few days earlier, reminding Calvert of the bitter disappointment recorded by Captain Scott on finding that Amundsen had forestalled him at the South Pole. The *Skate* became the second submarine to reach the North Pole on August 12 having cautiously surfaced in a patch of clear water on the way. These openings, or polynia, varying from cracks to small lakes are caused by the action of the Arctic winds and currents on the pack ice, and Commander Calvert's orders had stressed the importance of developing techniques for making use of them. A few days after reaching the Pole the *Skate* surfaced in an opening close to Station Alpha, a camp of scientists engaged on work for the International Geophysical Year. The station was moving with the drifting pack ice and the scientists were uncertain of their position, but, by arrangement, noise of a motor boat engine enabled the *Skate*'s sonar to home on to the spot. Her visit to Station Alpha was, however, cut short by ice closing in on the submarine and after hurried farewells, the *Skate* dived to safely. When she finally emerged from the ice pack on August 20, she had been under the ice for 11 days and had travelled 2,405 miles. More important still, she had demonstrated on nine separate occasions, two of them close to the Pole itself, that a nuclear submarine could surface in the polar ice pack.

The next step was to demonstrate that a nuclear submarine could operate in the Arctic in mid-winter : a very different proposition with temperatures some 60 degrees below the summer average accompanied by gale force winds. This time a television camera was fitted for viewing the underside of the ice and the sail strengthened to break through the thicker ice likely to be encountered in the frozen lakes. On March 17, 1959 the *Skate* made history by breaking through the ice at the geographical North Pole where a funeral service was held for Sir

Hubert Wilkins, who before his death the previous December, had asked that his ashes should be deposited at the Pole by one of the nuclear submarines.

Her voyage completed, the *Skate* had travelled 3,090 miles under the winter ice and surfaced ten times: a convincing proof that it was possible to travel freely under the ice at any time of the year and use the openings or thinner patches of ice to surface reasonably close to any pre-arranged spot – a capability of great significance for naval warfare in Northern waters, for missile-armed submarines could do likewise.

Another submarine of the *Skate* class, the *Sargo*, became the third vessel to reach the North Pole, having made a winter passage through the Bering Strait and the shallow Chukchi Sea, where the distance between ice and sea bed averages only 150 feet – not much room for manoeuvre in a nuclear submarine which measures 50 feet from fin to keel: moreover, at this time of the year the ice pack stretches over 1,000 miles further south than in summer. During her voyage, the *Sargo* surfaced 20 times in the ice pack, at times breaking through ice four feet thick.

The fourth submarine of the class, the *Seadragon* made a particularly valuable contribution to Arctic operations, that of completing the first submerged transit of the famous Northwest passage, between Canada and the frozen Arctic, and by the most direct channel. This would make it easier in the future for nuclear-powered submarines to pass from the Atlantic to the Pacific.

A further advance was made when the *Seadragon* and *Skate* met under the ice in July 1962, and together headed for the North Pole, where they both surfaced. This was the first multi-submarine operation in the Arctic and the exercises they carried out added considerably to the knowledge of naval warfare conditions in Northern waters.

The earlier nuclear submarines, including the *Skate* class had the hull form of conventional vessels, notably that of the *Tang* class of post war attack submarines, which had developed from the wartime fleet type of basically 'ship' form. The high underwater speeds introduced with nuclear power – the *Nautilus* was unofficially reported as having a submerged speed of 23 knots – called for a more streamlined hull form that would give a maximum underwater performance. It was decided to build a small, fast, diesel-electric submarine, the *Albacore*, to determine the best hydro-dynamic shape. Results were firmly in favour of a short, fat streamlined body with a tapering stern and single propellor : a 'throw-back' to the original design of the inventor John P. Holland.

The union of the streamlined *Albacore* hull and nuclear power resulted in the *Skipjack*, whose performance on trials was sensational. Officially described as the fastest submarine in history, the US Navy would only admit to a speed 'in excess of 20 knots' : unofficial estimates credited her with 35. Although 15ft shorter than the *Skate*, her large-diameter hull gave her considerably more displacement. Launched in May 1958 the *Skipjack* has been described by Captain L. Meach USN, an experienced submariner, as 'unsurpassed in speed, manoeuvrability, fighting power and overall excellence'. All subsequent American and British designs of both *Polaris* and fleet submarine types, have been derived from it.

British submarine designers had been closely watching American developments and an early intimation appeared in the Naval Estimates of February 1952 : 'All possible means of submarine propulsion are under investigation, including systems using nuclear energy'. But seven years were to pass before this was translated into action : a long time for a nation that had been in the forefront of nuclear development. The answer appears to be in the priority being given at that time to nuclear missiles, which left insufficient fissionable material available for marine use. Due to the foresight and drive of Admiral Earl Mountbatten, who became First Sea

Lord in 1955, this policy was changed. He at once sent an observer to Washington; factories building reactors for the US Fleet were visited, and in December 1957 an agreement was reached with the United States Government, part of which sanctioned the purchase of a complete set of propulsion machinery of the type fitted in the *Skipjack* : the submarine herself was to be British built.

The *Dreadnought* was laid down at Vickers, Barrow-in-Furness in June 1959, and, contrary to the usual British practice, the vessel's name was disclosed from the start instead of shortly before launch. The name was a famous one, recalling Admiral Fisher's revolutionary new battleship, completed in 1906, which first introduced the all big-gun main armament and steam turbine propulsion for large warships. Mr Rowland Baker, then Assistant Director of Naval Construction, who had just returned from loan duty in Canada was put in charge of the technical side. Basically of *Skipjack* design, the *Dreadnought*'s bow was more blunt and whale-shaped and her hydroplanes were fitted in the traditional bow position whereas the *Skipjack*'s were mounted on the fin. But from the fin aft she was identical with *Skipjack* and had a single propellor shaft : like the *Skipjack* she carried six bow torpedo tubes.

Meanwhile, the crew were given the necessary training, Commander B. F. P. Samborne, the commanding officer designate took a course of nuclear science and technology at the RN College, Greenwich, followed by eight months training in the USS *Skipjack*. Other key officers and ratings were trained on board American nuclear submarines and in shore establishments in the USA. The *Dreadnought* was launched by the Queen on October 21, 1960 and commissioned on April 17, 1963. On joining the Fleet after a working up period she soon made her mark and opened the eyes of naval experts to the possibilities of the nuclear submarine.

An aspect of submarine warfare that had been uppermost in the minds of British and American planners for some time was the anti-submarine submarine. The idea was not new. During the 1914-18 war, British submarines sank 17 German U-boats and towards the end of the war a special type of 'R' class submarine was built with a high submerged speed specifically for anti-submarine work. Displacing 410 tons, they had a single screw – a rarity in those days – and a new arrangement of twin rudders which made them extremely manoeuvrable: they also carried six bow torpedo tubes and advanced hydrophone gear. While their surface speed was only nine knots, electric drive from large batteries gave them an underwater speed of 15 knots. But this promising development came too late to have any effect on the war, and the majority of them were scrapped soon after completion. Henceforth the main anti-submarine effort was concentrated on improving the detection equipment and weapons carried in surface ships – until the advent of the nuclear submarine.

The huge Russian submarine fleet of 350 vessels was an obvious threat to the free world: Hitler had started the 1939 war with 57 and they had done damage enough. Thus, interest in the anti-submarine submarine revived, and its role came to be described as 'hunter-killer'. The US Navy decided to build a quiet, extremely manoeuvrable submarine, the *Tullibee*, with the accent on sonar – her entire forward compartment was devoted to it, her four torpedo tubes being mounted amidships. To eliminate noisy reduction gearing she was given turbo-electric drive, as opposed to the steam turbines in other nuclear submarines. Some 20 foot longer than the *Skipjack*, the *Tullibee* displaced 2,318 tons. Commissioned in November 1960 she was the only one of her kind owing to the success of the *Thresher* class, which incorporated the best features of the *Skipjack* and the *Tullibee*.

To understand fully the development of this type of submarine, it is necessary to examine the nature of the submarine versus submarine encounter. Taking place perhaps hundreds of feet below the surface, the two protagonists are pitting supreme professional skills against each other : it is single combat *par excellence*, akin to the clash between mediaeval knights. Differing in almost every respect from the encounters between other forces – land, sea and air, the submarine has to rely exclusively on information obtained by sound. Compared with visual or electronically obtained information this is slow, intermittent and incomplete. Nor can any concentration of force be brought to bear which would achieve tactical success over the enemy : it is a hand to hand combat fought to the death. Victory goes to the one that first detects, correctly classifies and pinpoints the other's position. Much depends on the respective noise levels as the two submarines approach each other and the advantage may well lie with the vessel on patrol quietly lying in wait while its adversary, hurriedly and noisily on passage, passes unsuspectingly by. Much depends on the efficacy of the detecting equipment and the homing torpedoes or other weapons used.

Embodying the latest equipment for anti-submarine warfare, the *Thresher*, name ship of a new class, with a *Skipjack* hull and power plant, was given amidships torpedo tubes and a sonar-packed bow section. Improved steels were used in her construction which would enable her to dive deeper and withstand greater pressures than any previous submarine. She was also equipped with Subroc, a new high performance anti-submarine missile, capable of carrying a nuclear warhead. Fired from a torpedo tube this weapon surfaces and takes to the air, re-entering the water in the vicinity of the enemy vessel.

The *Thresher* was laid down in May 1958 at the Portsmouth Navy Yard in New Hampshire : unlike the first of class of all previous nuclear submarines which had been built commercially by Electric Boat of Groton, Connecticut. Commissioned three years later, the *Thresher*'s career was brief and tragic. After a year of tests, trials and exercises she was taken in hand for refit at the Portsmouth Navy Yard where she remained for some nine months.

During the refit, a new captain was appointed, Lieutenant-Commander John W. Harvey USA, an experienced officer who had served in the *Nautilus* during her polar voyage, later in the *Tullibee*, and more recently as executive officer in the *Seadragon* during her famous meeting with the *Skate* at the North Pole.

On the morning of April 9, 1963 the *Thresher* put to sea for post-refit trials accompanied by the submarine rescue ship *Skylark*, commanded by Lieutenant-Commander Stanley Hecker USN. On board were a number of dockyard officials and firms' representatives, making a total of 129. The *Skylark* was able to communicate with the submarine by underwater telephone, but the diving bell she carried could only attempt a rescue if the *Thresher* were in difficulties over the continental shelf, where the depth did not exceed 600 feet. The shallow diving tests that day were successfully concluded, and a rendezvous between the two vessels was arranged for the following day some 200 miles East of Cape Cod, where deep diving trials could take place clear of the continental shelf. The next morning the submarine reported that she was starting a dive to her test depth – a maximum operating depth that has never been officially released but has been unofficially estimated as between 800 and 1,000 feet.

At 7.52 a.m. she had reached 400 feet and the dive was temporarily stopped for a few minute for a routine inspection for leaks. At 8.09 a.m. the *Thresher* reported that she was at one half test depth and from then on all references to depth were reported in terms of test depth, for security reasons. At 8.34 she was at test depth minus 300 feet and at 8.53 was 'proceeding to test depth'.

At 9.12 there was a routine communication check between the two ships. A minute later a somewhat distorted message was received by the *Skylark* – the underwater telephone is seldom very clear – and those who heard it have variously reported it. Lieutenant-Commander Hecker heard : 'Experiencing minor problem . . . have positive angle . . . attempting to blow'. This was followed by a sound resembling air under pressure being blown into the ballast tanks. At 9.17 a final message reached the surface which Hecker found too garbled but his navigating officer was convinced it contained the words 'test depth' followed by sounds of a vessel breaking up. The depth of water in this area is 8,400 feet : long before reaching that depth the *Thresher*'s hull would have been crushed by the enormous pressure.

An intensive search began for any clues as to the cause of the disaster. Any danger from radioactivity was discounted by a statement on April 11 from Vice-Admiral Rickover 'Reactors of the type used on the *Thresher*, as well as in all our nuclear submarines and surface ships, can remain indefinitely in sea water without creating any hazard'. A Court of Inquiry was immediately convened, but its findings were not made public till June 20. These stated that a flooding casualty in the engine room was the 'most probable' cause of the sinking : 'the Navy believes it most likely that a piping system failure had occurred in one of the *Thresher*'s salt water systems'.

Meanwhile, the search for wreckage continued and when underwater photographs revealed some likely looking debris, the Navy decided to employ the only craft capable of operating at such depths, the 50 ton bathyscaphe *Trieste*, which had been bought from her designer and builder Professor Auguste Piccard in 1958 and was based at San Diego, California. This remarkable vessel had already in January 1960 achieved the world's deepest dive of 35,800 feet in the Challenger Deep in the Pacific. Transported by ship via the Panama

Canal, the *Trieste* was soon in operation. A more detailed account of the *Trieste*'s search will be found in Chapter 9.

Only one nuclear submarine has been built with two reactors, the radar picket *Triton*, which entered service in November 1959. Her powerful engines gave her a surface speed of over 30 knots, to enable her to operate with fast carrier striking forces : for this reason she was given a ship-type hull. The intention was that submarines of this type would be stationed in a circle around an amphibious force to give warning of air attack, after which the submarines would submerge. But the use of submarines for this purpose was discontinued when more powerful radar for surface ships and aircraft was introduced, and in 1960 the *Triton* was given the task of a round-the-world voyage submerged – her most memorable feat. The overall submerged distance she covered was 36,000 miles taking just under 84 days, but the actual circumnavigation, which began and ended at St Paul's Rocks in mid-Atlantic and involved rounding the Horn, was achieved in 60 days. The impact of the voyage was somewhat dampened by the U-2 spy plane incident shortly before the submarine's homecoming on May 11, but Mr William B. Franks, Secretary of the Navy announced that the *Triton* had demonstrated that American submarines 'can go anywhere they wish and for as long as they wish'. The following year, the *Triton* was reclassified as an attack submarine.

Another nuclear submarine associated with a development that was short-lived was the USS *Halibut*, designed to carry the 600 mile cruise missile Regulus. Apart from her forward compartments which had to accommodate a large hangar for five 57 foot winged missiles, not unlike the German V-1 'buzz bomb' launched against Britain in 1944, the submarine resembled the *Skate*. A serious drawback was that the submarine had to surface before firing her missiles, and it is not surprising that with the rapid development of the Polaris system, the requirement for the *Halibut*-type vessel ceased and on the strength of her nuclear power and the torpedo armament she was reclassified as an attack submarine.

Meanwhile Britain's own nuclear submarine programme was forging ahead. Mention has been made of the December 1957 agreement, which amongst other things, enabled Britain to purchase enriched uranium from America under specially favourable conditions. Admiral Rickover did much to further the co-operation, partly from his respect for British naval experience and partly on account of the help he had himself received from the Royal Navy in the early days of the war, which had enabled him to set up an efficient repair service for the US Fleet.

A prototype submarine reactor was built at Dounreay, Ross-shire in the spring of 1958 for use in experimental work and for training the crews to man Britain's nuclear submarines. A replica of part of a submarine hull, the prototype containing the reactor, propulsion machinery and control panels was immersed in a large seawater tank. With a core made by Rolls-Royce the reactor went critical for the first time early in 1965.

The first all-British nuclear submarine, HMS *Valiant*, was launched at Vickers, Barrow-in-Furness in December 1963. Her commissioning, delayed for several months owing to the prior claims of the Polaris programme, took place in July 1966. Twenty feet longer than the *Dreadnought* and displacing 500 tons more, the *Valiant* was a success, and in April 1967 completed the homeward voyage of 12,000 miles from Singapore, the record submerged run for a British submarine, in 28 days.

Also of the *Valiant* class are *Warspite*, *Churchill*, *Conqueror* and *Courageous*, while three vessels of an improved *Swiftsure* type have been laid down and a fourth is on order. The Navy hopes eventually to have a force of about 25.

The distinction of being the first ship in the Royal

Navy to have reached the North Pole fell to the *Dreadnought* on March 3, 1971 when she surfaced through ice a foot thick. The round voyage of 5,200 miles had taken her 19 days in 'the worst time of year from the point of view of the temperature and daylight' said her captain, Commander Alan Kennedy.

The experimental work at Dounreay with reactor cores of Rolls-Royce design has borne fruit: the *Swiftsure* should be able to operate for twice as long between refuelling as the earlier *Valiant*.

CHAPTER FOUR

The Patrol Submarine

Towards the end of the 1939-45 war, the Germans introduced a highly successful type of conventional submarine, the Type XXI. Displacing 1,600 tons, with a submerged speed of 17 knots and 14 knots on the surface, they had a remarkable endurance which enabled them to patrol for three weeks off Cape Town from Germany and return without refuelling. Furthermore they could dive to 850 feet and were almost noiseless at slow speeds. They carried six torpedo tubes and rapid reloading gear with an outfit of 23 torpedoes.

These vessels were pre-fabricated, their hulls being built in sections at inland factories and assembled at shipyards. Although a number of them were completed, it was perhaps fortunate for the Allies that only one of them, the U-235, put to sea a few days before the surrender, and she achieved no more than a dummy attack on an unescorted British cruiser after the order to cease operations.

Faced with a large number of these vessels, even Britain's efficient anti-submarine escorts would have been hard put to it, mainly on account of the type XXI's high underwater speed. Able in coastal waters to find the convoys without surfacing and close them to firing range, they could keep up with the convoy submerged and fire three salvoes in rapid succession, afterwards making a quick get-away. In this way, they could be expected to inflict four times as much damage to a convoy as existing U-boats. In basic terms it meant a race between the Type XXI and the vast American shipbuilding effort.

At the Potsdam Conference after the war, Churchill,

Truman and Stalin met to divide the spoils of war, including the surviving U-boats. It was finally decided that the larger part of the German submarine fleet should be sunk and that not more than 30 submarines would be preserved and divided equally between America, Britain and Russia 'for experimental and technical purposes'. After the conference, 114 of the surrendered U-boats were scuttled at sea in deep water. In addition to her official quota, Russia succeeded in shipping back to the homeland several uncompleted U-boats found on the building ways and fitting out basins at Stettin and Danzig, some of which were completed and put into service. A number of submarine experts captured by Russian forces went to the Soviet Union.

As a result of trials with the Type XXI, America and Britain began a programme of modernizing existing submarines and designing new types of their own. The Americans scrapped over 70 of their older submarines, cancelled another 92 of those building, leaving just under 200 of the larger ocean-going fleet type.

Many of these were given a 'Guppy' conversion (Greater Underwater Propulsion Programme) to improve their underwater speed, which included streamlining of the hull and superstructure on German lines. This was followed by a new design of *Tang* class – the American version of the type XXI.

Britain scrapped 45 old and worn out boats and cancelled another 50 on the stocks, leaving a fleet of just under 100. The existing 'T' class was lengthened and streamlined, whilst 16 of a new 'A' class were completed. Originally designed for the Pacific war, with a high surface speed of 19 knots and eight knots submerged, the 'A's were all welded vessels with a different hull from the 'T' class : many of them carried a 4 inch gun forward of the conning tower. Schnorkels or 'snorts' were fitted to the entire British submarine fleet, greatly improving their underwater performance.

The first British post-war submarine design based on German ideas, was the *Porpoise*, eight of her class being completed between 1955 and 1961. Displacing 1,605 tons, they had a submerged speed of 17 knots and 12 knots on the surface, six bow and two stern torpedo tubes and a complement of 6 officers and 65 men. Designed to be very silent they were also given greatly improved acoustic detection gear, to counter enemy submarines : now the primary operational task of British submarines.

During the war, the Royal Navy's submarines had sunk 40 German, Italian and Japanese U-boats, all of them on the surface, except for two, almost at the end of the war which had been submerged at the time. The first submerged underwater sinking was in February 1945 off Norway, when HMS *Venturer,* commanded by Lieutenant J. S. Launders, picked up on her hydrophones the sound of another submarine's engines. A quick periscope search revealed another periscope nearby. The *Venturer* stalked the U-boat for over two hours until the two craft were 2,000 yards apart, when the British vessel fired four torpedoes on the hydrophone bearing, at least one of which scored a hit. When she later surfaced, the *Venturer* found a quantity of wreckage from *U-864*. Most of the U-boats destroyed by allied submarines during the war were sunk in enemy waters, and the original intention was to use the new patrol submarines offensively in the same way.

The main impetus in allotting a primary anti-submarine role to American and British submarines was the rapid growth of the Russian submarine fleet, which had ended the war with between 100 and 200 vessels and subsequently built up its numbers until by 1950 there were about 350, the latest of which were assumed to embody recent German developments. The nature and *raison d'etre* of this formidable force will be dealt with in a later chapter. While the Russians were increasing their submarine fleet, British and American anti-submarine forces against improved sonar equipment and homing torpedoes which would seek out the enemy at any depth.

Following closely on the *Porpoise* class came the *Oberons*, 13 of which were built for the Royal Navy between 1961 and 1967, the work being portioned out evenly between the principal submarine building firms, Vickers, Barrow-in-Furness; Cammell Laird, Birkenhead; Scotts, Greenwich; and Chatham Dockyard. With improved detection equipment, a high underwater speed and remarkably silent, the class represents the ultimate in conventional submarine design, and embodies also for the first time, glass fibre in the superstructure before and abaft the conning tower. In common with all British submarines, they are fitted with high definition radar, mainly as an aid to navigation in confined waters or poor visibility. Interception equipment enables them to detect the searching radar from enemy ship and aircraft, giving them time to avoid detection by deep diving or other evasive measures.

Patrol submarines, which will form the bulk of Britain's submarine fleet for some years to come, still have an important role to play. Their ability to remain submerged for several weeks, their high underwater speed, well-proved and efficient control systems, up-to-date sonar and above all their silence, make them formidable adversaries, able to perform a number of roles.

The notion of patrol is derived from the French *patrouiller*, to go the rounds, like a guard, whose duty is to march round a camp during the night and see to its safety. It implies also a presence ahead of a main force for reconnaissance or the testing of enemy reactions. It can cover both offensive and defensive operations of a minor nature. Applied to submarines this can mean a variety of tasks, of which the most important is attack on enemy warships and supply vessels. They can reconnoitre beaches before an amphibious attack, carry out clandestine missions such as the landing of agents or saboteurs, take special cargoes to beleagured garrisons, or make small raids on heavily defended key points. There is also a continual training task for which nuclear-powered submarines are uneconomic or not otherwise suitable. Not only must future commanding officers and crews be trained, but the Navy's own anti-submarine forces; new tactics must be developed, new detection systems and anti-submarine weapons evaluated. The French, the Dutch and the West Germans have continued to develop the patrol submarine, while the Soviet Navy, which has so rapidly expanded its nuclear submarine fleet still appears to need a small number of new conventional vessels, presumably for training.

The question arises whether on a limited budget it is better to have a small number of highly sophisticated vessels or a larger number of simpler and cheaper ones. In other words, should there be fewer Fleet and more Patrol submarines? When between four and five Patrol submarines at £6 million each can be built for the cost of one nuclear-powered Fleet submarine, it is small wonder that even the experts differ. A closer look at the arguments will underline some of the main characteristics of the two types.

In a direct confrontation between Fleet and Patrol submarines, not all the advantage lies with the faster and more powerful Fleet type : much depends on how much noise they make. The vessel already on station in its patrol area, lying in wait, or cruising around in leisurely fashion, and well nigh silent, has a distinct advantage over one 'in transit', proceeding at speed with an inevitable build-up of noise. Again, shallow water to some extent limits the operation of the larger nuclear Fleet submarine, giving the Patrol submarine an additional advantage. Naval bases are generally situated in comparatively shallow water, so that a Patrol submarine on station off an enemy base can present a serious and abiding threat.

On the other hand, the Fleet submarine has more powerful active sonar equipment with which to gain a tactical advantage.

It can greatly augment the anti-submarine forces and be decisive in countering torpedo or short range missile attacks on a convoy or task force, thus providing the strongest possible escort.

Patrol submarines suffer from the disadvantage that for much of the time they must expose their 'snort' apparatus, especially when on transit to their patrol area. While they are doing so they are liable to detection by maritime aircraft, underwater listening arrays, or, most dangerous of all, enemy submarines equipped for the anti-submarine role. Once at the patrol area, they can expect even more intensified anti-submarine measures, possibly from maritime aircraft alerted by information from acoustic apparatus on the sea bed. But the submarine will not easily be driven from its patrol by maritime aircraft if its snorting activities only take place every four or five days : more dangerous in those conditions are the attentions of surface hunting craft, supported by helicopters. If the submarine can be given not only effective anti-ship weapons, but a good counter to the helicopter, its power of survival will be considerably increased. The helicopter, equipped with 'dipping sonar' has to hover for a period some 30 feet above the water. This is the moment when it is especially vulnerable to the 'Blowpipe' type missile, now under trial for naval use, as the submarine-launched air missile (SLAM).

The patrol submarine can still play a large part in the protection of the Western Alliance's trade – principally by its inshore anti-submarine patrols. Russia's huge submarine fleet makes it possible that merchant ship losses would initially be twice as heavy as at the worst period of the 1939-45 war, when ten merchant ships were sunk for every one U-boat. The counter to this must come both from fleet submarines in an escort role and patrol submarines attacking enemy U-boats at source.

Britain's existing force of patrol submarines has a life span until about 1985. Are we to build more to succeed them? When other European navies, and even the Russians, are still building conventionals, can Britain afford to ignore them? Private enterprise thinks not: Vickers, Britain's most experienced submarine builders are collaborating with the Kiel firm of Howaltswerke in producing 500 and 1,000 ton submarines designed by Professor Gabler of Lübeck. One of the larger of these was recently delivered to Greece, others are under construction at Kiel. Four of the smaller type are being built for Turkey.

The American Navy, on the other hand, has decided to build no more conventional submarines, the last non-nuclear combatant submarine, the *Bonefish*, was commissioned in July 1959. The few conventional submarines built by the Russians are thought to be for training purposes, and not a new class for operational service.

The Royal Navy with its still considerable overseas commitments and an extensive training task has not enough submarines to go round. The older modernized 'A' class is rapidly disappearing. There are now only three of them in service, *Andrew*, *Auriga* and *Alliance*: by the mid '70s these too will have gone. The training and routine commitments would then fall on the 21 hulls of the *Oberon* and *Porpoise* classes, providing a strong temptation to forward planners to call for the building of some interim diesel-electric boats for training, four of which could be built for the cost of one nuclear fleet submarine. The Navy has not adopted this solution, the main reason being that in today's political and economic climate it would inevitably mean curtailing still further the production of fleet submarines, whose building rate is down to a mere one every 15 months, as against America's four-and-a-half a year and Russia's 12 to 15 a year.

Given a life span of 15 to 20 years it must also be considered how the diesel-electric submarine, built in the near future would fare in the last decade of the century, with its limited submerged endurance and comparatively slow underwater speed. Anti-submarine measures, too, would be more deadly by then, while the patrol submarine's own traditional weapon against the surface ship – the torpedo – will probably have disappeared.

To put a new generation of costly and complicated weapons into ageing hulls would be putting new wine into old bottles. Much will depend on the development of the new submerged fire anti-ship weapon. If this can be fitted into conventional hulls without much difficulty and fired from existing torpedo tubes, there is a case for converting existing craft and perhaps building a new class of conventional submarines.

The submarine's training task will also change, since torpedo attack at close range will have given place to missile attack from beyond the reach of the hunting ship's sonar. Both sonar detection and delivery of the anti-submarine weapon will more and more become the special province of the helicopter. This again renders less vital the patrol submarine's contribution to the training of surface forces.

Overriding all other considerations is the fact that the performance of the nuclear submarine is so vastly superior to the conventional that the limited resources available are best applied to the nuclear vessel. The high speed of the nuclear submarine, exceeding that of a fast frigate or destroyer, can be sustained for almost unlimited periods and in any weather. Although the average speed of merchant ships has increased over the years by a few knots, the speed of the nuclear submarine gives it a greater margin over them than a surfaced submarine had in the 1939-45 war : it can overtake or intercept merchant ships or convoys more easily and get away quicker after an attack : ships or aircraft that arrive even a short time afterwards are unable to find it. The nuclear submarine's speed on passage being much greater, it can remain longer in the operational area to be redeployed to escort a task force or counter an enemy attack.

The endurance of the nuclear submarine is virtually unlimited and far greater than the largest conventionally-powered vessel afloat, as was shown by the US submarine *Triton*'s submerged voyage around the world in 1960. This ability to cruise at speed in any ocean of the world is linked with a complete independence of bases, supply ships or tankers.

CHAPTER FIVE

The Russian Challenge

The Russian Navy, now the second largest in the world, has some 400 submarines, 100 of which are nuclear-powered – the largest submarine fleet the world has ever seen. It is building nuclear submarines, a growing number of which are fitted with Polaris-type ballistic missiles, at the rate of 12 to 15 a year.

This is only part of a remarkable change that has taken place in recent years : the emergence of Russia as a major maritime power. To appreciate fully the importance given by Russia to her submarine arm, it is necessary to view it against a whole new background of Russian naval thinking.

Until the beginning of the 1939-45 war, Russia as a great land power regarded its navy solely as a useful support for land operations : an extension of the army's flanks. Over the past 30 years the Russian Navy has not only built up a position second only to the United States in ships and man power, but has become a world-wide ocean going force. This change has been a gradual one : at times moving fast, at other times more slowly. Some of the more important milestones should be noted.

After the October revolution and the civil war that followed, little money was available for the rebuilding of dockyards and the laying down of new ships.

So the Russian Navy was limited to modernizing its few remaining ageing vessels. Any prospect of achieving maritime superiority except in certain inshore areas was out of the question. At the same time, new guiding principles were afoot derived from the guerilla warfare of the revolution, and the Navy was officially advised to develop strategy and tactics based on Marxist-Leninist ideas. As a result, a new school of strategists emerged which decried traditional British and American naval doctrine and the composition of fleets these implied. The submarine, they argued, had replaced the battleship as the main striking unit of the fleet : submarines, supported by aircraft and light surface craft, should form the basis of Russia's future navy. The submarine, it should be remembered, is the guerilla *par excellence*. This doctrine fitted in well with a number of other factors; it was inexpensive compared with a traditional big ship construction programme; it harmonized with the 'unified command' principle – a Leninist ideal – and it was suitable for the short coastlines of Baltic and Black Sea areas.

Stalin, who had basically accepted the small ship navy idea, became concerned in the mid-1930s with the growing Japanese threat in the Pacific and the increasing power of the German Navy in the Atlantic.

He decided on an expanded shipbuilding programme, to include battleships and heavy cruisers : this would, he thought, add weight to his diplomatic efforts and act as a deterrent to these new naval powers. At the outbreak of the 1939-45 war, the Soviet Navy had laid down over 500 warships, and those in service included four cruisers, some 40 destroyers and over 200 submarines, while under construction were two battleships, ten cruisers and a further 100 submarines. But the time-honoured notion of the Navy as a defensive force, confined to the protection of the country's coastline and the army's flank, persisted.

For a force capable of more protracted operations, the Russian Navy's contribution to the war was insignificant. The forces in the Baltic and Black Sea never left those areas, and those of the Northern Fleet were seldom seen by the great allied convoys carrying aircraft and tanks to Murmansk. In the Black Sea, the advancing German army destroyed the shipyards and some major units under construction, but the large ships remaining were not active, and their crews were used to man the increasing number of torpedo boats which often successfully worked in support of the Red Army offensive. It was here that a young Rear-Admiral Sergei Gorshkov made his name. On one ocasion in the Sea of Azov, when it appeared necessary to launch a seaborne attack on the Germans, Gorshkov as naval commander was undeterred by the lack of amphibious craft : he pressed into service all the fishing boats in the area and used them as assault vessels.

In the Baltic, the larger ships were damaged in the early days of the war, destroyer operations were limited by fuel shortage, and the few submarine operations attempted were severely hampered by German minefields and suffered from lack of training and experience. It is a picture of a Navy which, despite the forward-looking strategy of the pre-war years and Stalin's additional shipbuilding programme, was still relegated to a subordinate and insignificant role.

In the post-war years, Russian strategic thinking was dominated by the fear that the capitalist powers would seek to destroy the communist bloc and in particular the Soviet Union itself. They had seen that their potential enemies were major maritime powers with a proved amphibious capability. Moreover, while the Soviet homeland was protected by 500 miles of buffer state satellites, her sea frontiers were exposed in three areas, the Pacific which was sparsely inhabited, the Black Sea, leading to the centres of industry and the Baltic, which lay beside her lines of communications. A particular threat was seen in the American carrier-borne aircraft, capable of delivering nuclear weapons. This led to a re-appraisal of the strategy of the 1930s : to which Stalin had himself given approval, resulting once more in a mainly static form of defence, with submarines forming the chief element, backed up by aircraft and surface units. German 'wolf pack' tactics would be employed, whereby the submarine's disadvantages of slow speed and short torpedo range were overcome by using them in greater numbers. Since reinforcement of the four main fleets, Northern, Baltic, Black Sea and Pacific, would not be practicable, each fleet would have to be independent.

To meet the threat of invading amphibious forces, two defence zones were planned: an outer one 500 miles from the coastline, defended by packs of submarines, and an inner, 150 miles out, controlled by aircraft, destroyers, torpedo boats and minefields. But to equip adequately all four fleets called for something like 600 submarines. Stalin accepted this policy and the forces needed to implement it, at the same time changing to the idea of a big ship navy with heavy cruisers and additional destroyers.

After Stalin's death in 1953, the maritime threat was reassessed. Khrushchev decided first that although an all-out nuclear war with the West was unlikely, it would be the only war for which Russia would prepare: conventional forces could be run down and the money saved applied to space programmes and aid to underdeveloped countries. From this time the only maritime threat seriously considered was one of nuclear strikes on Soviet territory from carrier-based aircraft and the counter to this became the primary role of the Soviet Navy. Defence was still left to submarines, destroyers and aircraft, but the main weapon was to be the anti-ship and anti-aircraft missile, nuclear-tipped for deterrence. This limited threat meant that the size and shape of the Soviet Navy could now be reduced and the big ship building programme was cancelled. A number of the conventional 'W' class submarines were converted to take Shaddock cruise missiles, with a range of 200 miles, and a new 'U' class was specially designed for the role. *Kotlin* class destroyers, the first ocean-going Soviet design, which first appeared in 1955, were also converted to carry missiles. The first of the *Kynda* class guided missile cruisers of 4,500 tons carrying two quadruple Shaddock mountings, was completed in 1962. This new pruned Soviet Navy represented a considerable saving in the defence budget.

One of the most far-sighted of all Khrushchev's achievements with regard to the Navy was the appointment in 1955 of the young Admiral Gorshkov as Commander-in-Chief, a post he holds to this day. Gorshkov may well have had doubts about the limited role still assigned to the Soviet Navy.

In the late 1950s, the maritime threat from the West was seen by the Russians to have increased in ways which had not been foreseen. In particular, the aircraft carrier force had much increased its radius of action.

Instead of strike aircraft operating close to the Russian coastline, and attacking mainly coastal targets, they could now reach the heart of the mainland after landing from the Norwegian Sea or the Eastern Mediterranean. The defence formulated in 1954 whereby carrier-borne strikes were countered by cruise missiles launched from destroyers and conventional submarines, under the cover of land-based aircraft was not enough.

A far more dangerous threat had developed in the late 1950s from the Polaris submarine, against which the Russians had no defence. Anti-submarine warfare had received little attention until now : long range sonar and anti-submarine weapons were lacking and the large submarine force was entirely committed to attacking surface ships : it had no anti-submarine role. The new threats from carrier-borne aircraft and Polaris submarines were taken seriously, and fresh assessments were made of the shape and size of the Soviet Navy. An attempt to catch up in the field of anti-submarine warfare can be seen in the intense espionage effort made against the Royal Navy's Underwater Detention Establishment at Portland, leading to the conviction of the Russian spy Lonsdale, the Krogers and the British civil servants, Houghton and Gee.

The best counter to both threats was seen to be the nuclear submarine. Fitted with long range cruise missiles it could attack the carriers operating far from the Russian coast, and, with improved sonar equipment, the Polaris submarines in their own environment. A proportion of the existing nuclear submarine force was converted to fulfil one or other role, until a new design appeared capable of both. Among surface ships, the helicopter cruisers *Moskva* and *Leningrad*, each capable of carrying 20 medium-sized anti-submarine helicopters, indicated a certain preoccupation with the Polaris threat, for the helicopters fitted with 'dipping' sonar had emerged as the most promising method so far of detecting and tracking nuclear submarines. It was also decided to turn the tables on America and pose a similar threat from Polaris type submarines – if suitable weapons could be developed, as well as the vessels to carry them.

The trials of the first Russian ballistic missile submarines coincided with the defence reappraisal of the late 1950s, but showed them to be far behind their American counterparts, both in the range of the weapon and its method of launching, which was still above water. New designs were put in hand for a Polaris type submarine that would enter service in the late 1960s.

Admiral Gorshkov's new Navy was beginning to take shape, based on the conviction that both surface and underwater units would have to operate for long periods at great distances from their bases and in areas that were potentially hostile. This was a big departure for a Navy restricted for decades to defensive zones around the Russian coastline. Doubtless he would have welcomed the inclusion of aircraft carriers in his fleet for air cover and reconnaissance in distant areas, but the long period that must elapse before such a vessel could be designed, built and put to sea and the necessary expertise acquired to operate it, was no solution to an immediate threat. Admiral Gorshkov may well have felt that this was something that should have been done years ago. In fact the Russians did approach the British Admiralty in 1947 as to the possibility of buying a British carrier second-hand : the request got nowhere. The Admiral could be sure of one thing : that any aircraft in the vicinity of his fleet could be considered hostile and could at once be engaged by the Goa surface-to-air missiles now being fitted in all his major warships.

By 1962 a number of *Kotlin* class destroyers had been fitted with anti-submarine torpedoes and anti-submarine rocket launchers and renamed the *Kildin* class, and the first of four *Kynda* class missile cruisers with two groups of four Shaddock missile launchers had entered service. A second generation of nuclear powered submarines, the E-class, fitted with Shaddock anti-ship missiles, intended as the main answer to the carrier, was now joining the fleet.

The Cuban missile crisis of 1962 marked the greatest turning point of all in Soviet Naval thinking. This was no defence of the homeland, it was Russian offensive weapons pointed at the heart of the United States. But without sufficient maritime power the scheme was foiled by simple and fundamental recourse by the United States to a naval blockade. In the face of the American Fleet the Russians found themselves impotent : rebuffed and humiliated they were forced to withdraw.

The year after Cuba, the introduction of new equipment in the Soviet Navy was accompanied by intensive training of officers and men in extended operations overseas. Russian warships began to appear in company for long periods in the Atlantic and in 1964 surface ships started to move into the Mediterranean where their submarines had already been operating. This can be seen both as a counter to the American Sixth Fleet and a means of advancing Soviet political influence in the area. The strong American reaction to the land-based Cuban missiles deployed so close to the American coastline strengthened the decision to place greater emphasis on the Soviet ballistic missile submarine programme. The American emphasis on seaborne strategic weapons, with steady production of Polaris submarines and constant improvements in missile range and flexibility, was to be matched by a similar Soviet strategic system aimed at retaining the deterrent balance. The Americans had also by the early 1960s improved the capabilities of their attack aircraft carriers, not only in long range nuclear strike but in air defence and anti-submarine warfare. All these developments reinforced the new Soviet naval policy.

The fall of Khrushchev at the end of 1964 heralded a change in emphasis in Soviet military doctrine, towards stronger and more flexible conventional forces. The idea that general nuclear war would inevitably follow from any conflict involving the Soviet Union was abandoned – a change that fitted in well with the re-birth of the Navy, its new ships and equipment and the experience gained in operations far from their home bases. Replenishment at sea with provisions, stores, ammunition and fuel, carried out for the most part in open anchorages, with techniques borrowed from British and American experience, became more and more a feature of Russian Fleet activities. The Arab–Israeli war of 1967 provided the Russians with the opportunity of acquiring limited naval and air facilities in the United Arab Republic and of increasing the number of ships in the Mediterranean. This number fluctuated from time to time and there were seasonal exchanges between Black Sea and Northern and Baltic Fleet units : generally the numbers in the Mediterranean were between 25 and 60 warships and auxiliaries, of which some 10 to 12 were submarines.

This number of ships, operating at a distance from home bases and without air cover cannot be said to constitute a serious military threat, but the impact of the Mediterranean penetration on world opinion was considerable – a fact that was not lost on the Brezhnev-Kosygin regime. Here was an economical way of furthering Russian political strategy, which if extended to other areas of the world, with similar reactions, could offer important advantages without risk. 'Our Navy', said Admiral Gorshkov in February 1970, 'together with the strategic missile forces has become the Supreme High Command's most important means of solving strategic tasks.'

The next area in which to test this policy was the Indian Ocean, important for the containment of China and the assertion of Soviet Russia as the leader of world communism. Other advantages were that it would diminish Western standing, promoting Soviet influence in its place, and offset the growing Chinese influence in Tanzania and Zambia. There was also a need to demonstrate support for India against China, albeit unobtrusively in view of relations with Pakistan; a move that China, still insignificant as a naval power, would be unable to contest. It is ironic that this Soviet move, which follows the pattern of decades of British 'showing the flag' in the area, should have been materially assisted by Britain's withdraw from East of Suez. It was a peacetime penetration for political purposes : not a display of naval might, but it rightly drew the attention of the world.

A naval occasion of a different kind occurred in 1970 when the Soviet Navy put to sea at maximum strength in every ocean in the world. It was Lenin's 100th anniversary, and the exercise was appropriately called Okean or Ocean. It was an impressive performance : the defensive fleet of 10 years before had been transformed into an offensive fleet of considerable diversity, power and sophistication : blue horizons had replaced the landlocked concepts of the Soviet past.

In the forefront of this new global role is the large Soviet Submarine Fleet, which since the decision made in the early 1960s to extend its deployment, has gained considerable experience in distant operations. Without overseas bases, this has been achieved by the use of support ships, giving the Soviet Navy the capability of carrying out world-wide submarine operations against shipping. This accords with the traditional Soviet view that the submarine is a primary unit of the Fleet, only the emphasis has shifted from the defensive to the offensive, a change of role much assisted by the nuclear submarine, capable of extended patrols and powerful attack. The crews of these vessels are gaining more and more experience and a capability for extended operations clearly exists.

It may be asked why the Russian Navy which claims to be a world-wide force should have such a preponderance of submarines at the expense of surface vessels as to make it an unbalanced force. An explanation sometimes advanced is that it was a costly blunder stemming from the theory that any future struggle with the Western powers would lead to a second 'Battle of the Atlantic' with a lengthy war of attrition against shipping. The Russians noted how close the Germans, who started the 1939-45 war with 57 U-boats, came to victory and were determined to avoid their mistake of starting the campaign with insufficient force. This theory maintains that the Russians failed to appreciate that nuclear weapons would make a long drawn out commercial war most unlikely.

The second theory, which has been mentioned earlier, was that the fleet was to be a main defence against the American carrier strike forces, which during the 1950s formed a part of the Western nuclear deterrent aimed at industrial targets, or submarine bases and shipyards. The fact that the Western deterrent was now wielded by Polaris forces meant that more and more Soviet submarines would have to be converted to a hunter-killer role.

A third theory is that the Russians foresaw the coming of the strategic nuclear stalemate and planned a large submarine fleet as a means of exerting political pressure in the West or of undertaking a limited war against enemy sea communications.

The sea undoubtedly offers opportunities for exerting pressure without serious risk. Every day 120 ships from overseas enter Western European ports to discharge over a million tons of cargo. Half a million tons of oil are carried round the Cape of Good Hope each day to keep the wheels of Western Europe turning. British merchant shipping still plies its trade on all the seas from China to Peru. Economic growth depends on the free movement of goods and the provision of services, the majority of which must be carried across the oceans of the world.

Russia and her satellites are rapidly increasing the size of their merchant shipping fleet, but the Soviet bloc is far less dependent on sea-borne trade than are the Western powers because of their central geographical position and excellent internal lines of communication. One long term aim may well be to undercut the shipping interests of the West and upset the delicate balance of the free world's economy. Another method might be the detention of ships using Soviet or satellite ports on the pretext of failure to comply with local regulations or the closure of large sea areas to commerce and fishing to avoid interference with naval exercises : this could even be enforced by the laying or declaration of minefields. Such measures would result in little more than diplomatic protests or appeals to the United Nations.

Another and more sinister method might be harassment on the high seas, or the sinking of merchantmen by 'anonymous' submarines : a form of guerilla warfare at sea for which a large world-wide submarine force would be admirably suited.

In the meantime Russian naval effort is largely concentrated on the production of nuclear submarines to carry out three main roles : the delivery of Polaris type missiles, the anti-submarine task and the anti-ship task :

general purpose submarines being built to combine the last two roles. Both the United States and Russia have close on 100 nuclear submarines, with the Red Navy having slightly more at this time. Current American building of nuclear submarines proceeds at $4\frac{1}{2}$ a year compared with 12 to 15 in Russia; since 1963 the Russians have sent at least six new nuclear submarine designs to sea compared to only two for the US Navy; American yards building nuclear submarines have decreased from seven to four, while Russian yards building these vessels have increased from two to four or possibly five. By 1975 it is estimated that Russia will have more ballistic missile submarines than America; moreover, according to Admiral Rickover, Russian nuclear submarines are improving in quality, especially in submerged high speed capability, endurance, quietness and weapon systems. There is every reason to believe that Soviet Russia is exploiting to the full the supreme advantages of the nuclear submarine.

CHAPTER SIX

The Submarine and the World's Navies

While America is rightly regarded as the true home of the submarine and Britain, Germany and Russia have, since the turn of the century, been prime movers in submarine development, their advances have been followed very closely by France, which from the earliest days has had a special interest in submarine warfare.

A major land power, needing a large army to secure her frontiers, France also possesses an Atlantic coastline and important maritime interests. To devote adequate money and resources to both land and sea elements was a strain on the national economy : a large battle fleet to match those of major naval powers being out of the question, other means had to be found of waging war at sea less expensively. One method was the *querre de course*, or the use of fast cruisers to attack trade. It was later realised that to circumvent an enemy blockade, submarines would be more effective and in 1886, Admiral Aube, the Minister of Marine, against much opposition, decided to adopt them. Submarines of various designs were tried, the most effective being the 260 ton electric drive *Gustave Zédé* built in 1893. Thus, the French Government was the first to sponsor official development of the submarine for strategic purposes.

In 1896, the French Minister of Marine organized a competition for a submersible which could operate on the surface as a form of torpedo boat but complete its attack submerged. Of the 29 designs sent in, the most promising was that of the *Narval*, with a dual propulsion system, steam and electricity. She had a durable hull, the inner being of submarine shape and the outer that of a torpedo boat. Her endurance was greater than that of the *Gustave Zédé*, which enabled her to operate anywhere in the Channel, but she had some serious drawbacks : she took 20 minutes to dive and the heat from her boiler made life almost unbearable when submerged.

Meanwhile the *Gustave Zédé* had proved in manoeuvres that she could torpedo warships at sea and in harbour. Her endurance, which enabled her to patrol all day within a radius of 30 miles from base, was clearly significant for defence. The French, sure in the knowledge that the submarine was now a viable warship, began to order in quantity vessels of both the *Gustave Zédé* and the *Narval*, or submersible type. By 1900, with a total of 14 built or building, France was well in the lead of the six submarine navies.

During the 1914-1918 war four French submarines assisted nine British in the Dardanelles and Sea of Marmora in preventing seaborne supplies from reaching the Turkish Army in Gallipoli. The passage through the Dardanelles was hazardous, the submarines having to compete with a tortuous passage and strong currents as well as mines, nets, shore batteries, torpedo bombers and enemy patrols : over 50 per cent of the submarines engaged were casualties.

But the submarines had achieved much, not only sinking warships, transports and large numbers of small supply vessels, but shelling troops, bombarding railways and even blowing up a bridge by means of a landing party.

The inter-war years were marked by much international argument about the use of the submarine. The end of the 1914-18 war, with the surrender or destruction of the German U-boat force – the largest and most powerful the world had ever seen – left the British submarine fleet with 137 boats, more than any other nation. Despite this and the fact that Britain led the world in submarine development, the British Government believed that the submarine in the hands of an enemy was so dangerous that it outweighed any advantage in possessing it themselves and that total abolition of the submarine should be their aim : in short, a return to St Vincent's policy of 1803. This was the line taken by Britain at the Washington Conference of 1921, when the five largest naval powers assembled to avoid an armaments race and negotiate a treaty of limitation.

The British delegation maintained that the submarine was only effective as a commerce raider and then only if it disregarded international law, adding that it was of no use as a military weapon – a statement so patently untrue that it spoiled their case. Abolition was opposed by all the other countries at the conference, especially France, who had consistently maintained that submarines were essential for any navy with an inferior battle fleet: it was a means of redressing the balance. The French even demanded a submarine tonnage three times as large as the capital ship ratio would have given them. In this they were supported by Italy and Japan, both of whom had small battlefleets. Japan at that time had very few submarines, but her appetite had been whetted by eight ex-German U-boats, and she now had 12 of her own under construction.

The Washington Conference, while it did much to halt the armaments race as far as surface ships were concerned, was ineffective in limiting submarines. The most that Britain could do was to persuade the majority of members to sign a declaration that they would not use them in war for the wholesale destruction of commerce, as the Germans had done. But it was not at all clear whether this implied a total ban on commerce raiding by submarines, or whether it merely restricted them to the observing of international law in the matter of visit and search, as for surface ships. This part of the treaty the French refused to sign.

Having failed to secure the abolition of the submarine, Britain had to turn her attention to counter measures. Anti-torpedo bulges began to appear on the larger warships – inevitably with some sacrifice of speed. Considerable effort was put into the most promising detection device, the Asdic – to be described in greater detail later. Named after the Allied Submarine Detection Investigation Committee, formed in 1917, development work on this equipment was carried out at Portland, Dorset, in conditions of great secrecy. The outcome of the Washington Treaty had been that all countries except France would set out to develop the submarine as a military weapon only and not a commerce raider. Among new British submarine designs at this time was the *X-1*, the largest submarine yet built. Essentially a cruiser submarine, with four 5.2 in. guns as well as six torpedo tubes, she was capable of engaging enemy destroyers on the surface. It was a first step towards a somewhat bizarre notion of a fully submersible fleet, comprising ships of every type, which could disappear beneath the waves at will and pursue further activities underwater. But *X-1* was too slow to work with the Fleet, even on the surface, and was really best suited to act as a commerce raider – the very purpose Britain had forsworn. An imaginative design, which came near to being a success was that of *M-2*, one of three submarines originally armed with a single 12 in. gun – the 'Monitor' submarines. *M-2*'s big gun was removed and in its place was put a watertight hangar, large enough for a small seaplane. The submarine was able to surface, catapult her aircraft and dive again in the space of five minutes. But her tragic loss, with all hands in a flooding accident in 1930, put an end to the experiment.

One unlooked-for result of the Washington Treaty was the stimulus it gave to submarine building vis-à-vis the growing naval strength of Japan. Britain and America had agreed not to build new bases closer to Japan than Singapore or Pearl Harbour, while Japan had agreed not to open up bases overseas. This meant, of course, that in the early stages of a war in the Far East, Japan would enjoy command of the seas, and that, as the weaker powers in the area, Britain and America would be increasingly dependent on submarines.

Britain found that she needed a more suitable submarine for use in the Pacific than the 'L' class, her largest patrol submarine so far, developed during the 1914–18 war. This led to the *Oberon* class, ordered in 1923, with greater endurance and speed, as well as better habitability and diving qualities : they also carried six torpedo tubes. Nineteen of these vessels, of the 'O', 'P' and 'R' classes, which included one minelaying version, were built during the next ten years.

The Americans too, doubled the size of their existing submarines and produced a new 'V' class, largely based on the surrendered German cruiser submarine *U-140*. With two six-inch guns and displacing nearly 3,000 tons, they were reminiscent of the huge British *X-1*. In the event, these submarines too proved too large and unwieldy, and development centred on obtaining speed, endurance, habitability and a good torpedo armament in a slightly smaller submarine.

The Japanese, already imbued with the importance of submarines, began to build up their submarine arm in the hope that this would redress the balance against America. In the next ten years they gradually took the lead in the number of large ocean-going submarines in the Pacific.

The French began a steady submarine building programme after the Washington Treaty and, predictably, the vessels, of over 1,000 tons were admirably suited for commerce raiding. The large cruiser submarine *Surcouf*, laid down in 1926, armed with two 8 in guns and carrying a seaplane, looked like being a trump card for such operations.

Worse still, the Germans began secret preparations for a new submarine fleet. A cover operation was set up in Holland under the guise of an engineering firm, and two submarines from its designs were laid down in Spain and Finland, the latter country also providing facilities for German crews.

With six years still to go before the expiry of the Washington Treaty, the main naval powers met in London in 1930 to decide what was to follow it. Britain, still manfully advocating the abolition of the submarine, had by now halved the size of her submarine fleet : her 53 vessels placed her last but one among the five major naval powers. But the Royal Navy's submarines were modern; the wartime vessels and all but one of the steam-driven 'K' class had been scrapped; their armament was second to none; all carried the 21 in. torpedo and over half of them could fire a salvo of six. The diesel-engined *Thames* class, with very good handling and diving qualities, had a speed of 23 knots on the surface.

At the time of the London Conference, the Americans, with 81 submarines, had numerically the largest force, but only six were of the new long-range type. France had 66 vessels, of which 44 were of the sea-going patrol type, all modern; with another 41 building; she would soon head the list. Japan, with 72, had not only doubled her submarine fleet since the Washington Conference, but was the leading submarine power in the Pacific. Italy, with 46, mostly of the smaller types, came last among the Conference powers, but had 21 building or on order.

The London Conference did manage to achieve some limitation of the world's submarine fleets. No submarines were to be built in future over 2,000 tons, or with guns larger than 5.1 in. For those that already exceeded these limits, like the British *X-1*, the French *Surcouf*, America's

Argonaut and a few others, special provision was made. The size of Britain's submarine fleet at the time, 52,700 tons, was taken as the norm, and Britain, America and Japan agreed to limit themselves to this size. France and Italy would only agree not to increase the size of their existing fleets. The Treaty left intact the French submarine force of 82,000 tons, the largest and most modern in the world.

The ten years preceding the Second World War saw the re-birth of two large submarine fleets, those of Soviet Russia and Hitler's Germany. The growing challenge of the Soviet submarine fleet has already been referred to, and it will be recalled that by 1939 Russia had numerically the largest submarine fleet in the world, but its effectiveness was much reduced by being split between the four fleets of the Arctic, Baltic, Black Sea and the Far East. Many of these too were of the small patrol type, or built for costal defence.

In 1934, a year after Hitler came to power, 24 small U-boats of the type already tested in Finland and two of the type tried in Spain, were laid down : these could only be used in the Baltic and North Sea, but further expansion could be expected. The following year, Hitler repudiated the Treaty of Versailles, and an Anglo-German Naval Agreement was negotiated, whereby Germany was enabled to build up to 35 per cent of the tonnage of all types of warship allowed to Britain by the London Naval Treaty, and up to 45 per cent in submarines. It even included a special provision for Germany to build up to 100 per cent of British submarine tonnage should the situation warrant it. Germany also signed a Treaty known as the London Submarine Agreement, which undertook that merchant ships would only be sunk in war in accordance with international law, with their crews rescued and not left in open boats.

This was certainly appeasement, in line with the general policy of the day, but in signing the submarine agreement Britain was no doubt influenced by the belief that in the Asdic she had the complete answer to the submarine, and a rapid programme was started for equipping all the smaller ships of the fleet. Convinced that no other country had Asdics, Britain went ahead with her own submarine plans. The large *Thames* class submarines, no longer fast enough for a fleet role, were assigned the task of fast patrol submarines, and a new 'T' class was designed for work in the Pacific, chiefly remarkable for their ability to launch a salvo of ten torpedoes. The Americans also built a fine type of large submarine for use in the Pacific, with ten torpedo tubes, and with this class they outbuilt the Japanese.

The re-birth of Germany's U-boat fleet and its near success in the 1939-45 war was mainly attributable to the genius and vision of Doenitz, who in 1935, while still a Captain, was given command of the new service. He envisaged an out and out war against British commerce and sought to defeat its most powerful protection, the convoy system. His plan was to oppose the concentration of escorts round the convoy with a counter concentration of U-boats. The convoys would be found by air reconnaissance, or by the U-boats themselves spread out in scouting formation. He also set about the building of 500 ton Atlantic U-boats, ideally suited for wolf-pack tactics. The whole scheme called for a force of 300 submarines, 100 of which would be on station in the Atlantic, 100 on passage, and the remainder in harbour for rest or refit. It was all to be part of Hitler's naval expansion programme or 'Z Plan', by which he hoped to reach parity with Britain at sea by 1945: a plan that was to be overtaken by events. The development of Germany's outstanding wartime submarine design has already been described.

France's submarine force, which at the outbreak of the war had consisted of 80 vessels, with 22 building, had by 1946 been reduced to 14, three of which were British 'U' class submarines on loan. When the British vessels were returned the next year, they were replaced by ex-German U-boats. Four British 'S' class were later lent to France, one of which, the *Sybille*, was lost with all hands in September 1952, and the French submarine fleet still numbered 14. This meant no more than ten submarines operational at any one time – the lowest figure for the French Navy for 50 years. Moreover, the very mixed nature of the force made its operation and maintenance difficult. It was decided to replace the whole fleet, starting with the French units, with vessels that would embody the lessons learnt in the war. Like other major naval powers, the French strove to emulate the characteristics of the famous German Type XXI vessels in endurance, submerged speed, silence, armament and snort tubes. The result was the 1,500 ton, ocean-going *Narval* class, similar to the British *Porpoise*, the American *Tang* class and the Russian 'W' class. These were followed by four submarines of a smaller class, the 500-ton *Aréthuse*, specially designed for anti-submarine warfare, and finally by nine vessels of the high performance *Daphne* class, begun in 1957. These diesel-engined submarines have a displacement of 869 tons, a submerged speed of 16 knots, carry 12 torpedo tubes – eight bow and four stern – and have a crew of 45 officers and men. Eleven of them were built, of which two have since been lost : the *Minerve* in January 1968 in the Western Mediterranean, and the *Eurydice* in the same area in March 1970. These submarines have caught the interest of a number of Navies, including Spain and Portugal who are to have four each; Spain building hers at the Cartagena yard, with French assistance. South Africa has three and Pakistan originally had three, one of which was sunk by surface ships and helicopters in the Indo-Pakistan war.

South Africa had originally wanted three *Oberons* to be built in Britain, but when the arms embargo was applied by the Wilson Government, she was forced to place the order with France. India also wanted *Oberons*, but having failed to obtain suitable terms, purchased four 'F' class vessels from Russia.

When France decided to build her first nuclear submarine in 1958, she asked America for a reactor, but the request was refused on the basis of the MacMahon law, which prohibited the supply of enriched uranium for any military use; France thus had to fall back on her own resources and build a reactor of her own. In March 1960 it was decided to build a land-based prototype, similar in all respects to that of a submarine. The plant, built at Cadarache Nuclear Research Centre, was started up in August 1964 and, when power was increased, ran continuously for eight weeks, simulating a round-the-world voyage.

The size of France's nuclear deterrent force, or '*force de dissuasion*', has been set at four, or possibly five, vessels to be completed by the end of the 1970's. The first, *Le Redoutable*, was laid down at Cherbourg Naval Dockyard in March 1964 and entered service in 1971. She was followed by *Le Terrible*, launched in 1971 and *Le Foudroyant* at the end of that year. The decision to build a fourth, *L'Indomptable*, was officially announced in December 1967, and it is now known that there will be a fifth, *Le Tonnant*. The 16 Polaris-type, 14-ton M-1 missiles carried by the first two vessels have a range of about 1,900 miles, but improvements have been brought about, largely as a result of intensive work with the experimental submarine *Gynnote*, which exists for the sole purpose of testing missiles and equipment for the nuclear vessels, and the last three will have the 16-ton M-2 missile with a longer range. As yet, France has no nuclear-powered Fleet submarines, but these may be forthcoming, as we shall see later.

The Dutch, in their naval policies have perhaps been closer to the British than any other European nation, sharing a common outlook on many Continental problems while both countries have residual commitments in the Far East. In the 1939-45 war, many common burdens were shared by British and Dutch submarine forces. At the beginning of the Pacific War, the Dutch had 15 submarines based on Sourabaya in the Dutch East Indies. By that time, Britain, who had in peacetime kept some 18 submarines in the Far East in case of war with Japan, had withdrawn them all to the Mediterranean. As a result of Anglo-Dutch staff talks, roughly half the Dutch submarines were placed under British command at Singapore : these were used to patrol the South China Sea and the Gulf of Siam, and when the Japanese invaded Malaya, succeeded in sinking three transports and damaging four others, for the loss of three Dutch submarines. After the fall of Singapore, Dutch submarines operated from Ceylon and Australia. At a time when Britain's main submarine effort was in the Mediterranean, the Dutch had four submarines operating there.

The present Dutch submarine fleet consists of seven conventional vessels, two of which, of recent construction, have an underwater speed of 25 knots.

Holland would like to build a nuclear submarine of her own, and in June 1964 announced that she would do so. Four years later a Defence Note, conceding that the nuclear-powered vessel was the submarine of the future, stated that the defence funds available would not permit the ordering of one in the near future.

The Scandinavian countries, like Western Germany, have submarines that are mainly of the coastal type, suitable for operations in the Baltic or North Sea. Sweden, however, has a new construction programme for five long-range streamlined vessels of 1,125 tons submerged displacement.

Among the Mediterranean powers, Italy had long been interested in submarines – she had bought seven of Britain's older submarines in 1914. In 1930 she had 46, with a further 21 building or projected. In August 1940 Italian submarines began to come out of the Mediterranean to take part in the struggle in the Atlantic, and at one time they actually outnumbered the Germans in the area. But they were ineffective, mainly because they patrolled submerged for most of the time, waiting for targets to appear. The Germans found it impossible to cooperate with them and finally, the Italians were allocated an area of their own.

Since the war, Italy has relied mainly on ex-US Navy submarines, but has built a class of four small coastal vessels of 460 tons with hunter-killer characteristics. She also plans two ocean-going conventional submarines.

Greece eventually contributed eight submarines to the war against the Axis powers in the Mediterranean : they were used initially to attack the sea communications with the Italian army in Greece and Albania, and later in the Eastern Mediterranean helped in the destruction of Axis shipping. In the late 1960's Greece ordered four 1,000 ton submarines from Western Germany, to be built at the Howaldtswerke yard at Kiel. Germany experienced some difficulty in carrying out this contract, due to the restrictions on German submarine building. By the Brussels Treaty, Germany is precluded from building submarines of more than 450 tons displacement, except for six that must not exceed 1,000 tons. The matter was finally resolved by the vessels being considered as part of Germany's quota. Greece has in addition two ex-US submarines.

The Turkish submarine fleet is also being supplemented by German-built vessels. An agreement was reached in January 1970 whereby Germany would provide four 450 ton coastal type submarines, some to be built in Germany and the remainder in Turkey with German assistance. Turkey is also to receive three more American vessels in addition to the ten she already has.

Egypt has a total of 13 submarines supplied by the Soviet Union, 12 of them of the ocean-going 'R' and 'W' classes of slightly over 1,000 tons.

Two South American countries, Brazil and Chile, have recently turned to the British *Oberon* class submarine for replenishing their fleets, having previously relied on ex-US Navy vessels. The two submarines ordered by Brazil from Vickers at Barrow-in-Furness will be the first to be equipped with a new weapon control system TIOS, designed by the firm. Chile's two *Oberons* are being built by Scotts of Greenock.

It was natural that the Old Commonwealth countries, Australia and Canada, with their autonomous Navies, should acquire their own submarine arms. For many years a squadron of Royal Navy submarines has been based on Sydney for operations in Australian and Far Eastern waters. This has now ended with the formation of an all-Australian submarine squadron equipped with four *Oberon* submarines built by Scotts of Greenock. The first of these, the *Oxley*, was ordered in July 1964, *Otway*, *Ovens* and *Onslow* followed at yearly intervals, and all are now in Australia. It has since been officially announced that Britain is to build two more *Oberons* for Australia. But the formation of the Australian Squadron has not meant the end of British submarines in Far Eastern waters; one is to be permanently based there as part of the Five Power Commonwealth Force.

Canada has a submarine-building tradition of her own, having built a number of small submarines – notably the 'H' class – for Britain during the 1914-18 war, but her recent need for more sophisticated vessels has resulted in an order for three *Oberons* from Britain, all built at Chatham Dockyard.

The existence overseas of these British-designed submarines lends force to the notion, advocated by some senior naval officers, of a fully-integrated Commonwealth Submarine Force, in which many common duties could be shared and economies effected in the training of crews and in submarine maintenance.

The Chinese Communist Navy has the world's fourth

largest submarine fleet, mainly achieved with Soviet aid before the rift between the two countries. First formed in 1949, the Navy made slow progress at the start owing to a shortage of experienced officers and ratings, but in 1954 a programme of expansion began, leading to the Chinese Navy becoming the largest purely Asian navy in the world. The latest figures show that she has one Soviet 'G' class conventionally-powered missile submarine, with three missile tubes passing through her large conning tower; but there is no evidence that China has as yet any missiles for it. Reports that she is building three nuclear-powered submarines are unsubstantiated, but her 32 ocean-going submarines, ten older training vessels and three coastal submarines could, if efficiently manned, present a serious threat in the Far East.

Since the withdrawal of up to 2,000 Soviet advisers in 1960, there has been an inevitable slowing up in Chinese naval construction, and a drop in the standards of training. But the potential remains, for the Chinese not only made good seamen, but are traditional guerillas, and many aspects of submarine warfare must appeal to them. Should another Gorschkov arise in China, the balance of world naval power could be profoundly affected.

CHAPTER SEVEN

Submarine-borne Weapons

The Polaris system, the strategic weapon wielded by the submarine, was developed in the remarkably short period of five years. From July 1960, when the first Polaris submarine, the *George Washington*, successfully launched a missile, a gradually increasing force of Polaris submarines began their peacetime patrols.

The next advance, initiated by the American Strategic Project Office under Rear-Admiral Levering Smith, was the introduction of Poseidon. With approximately the same range as the Polaris A-3 missile – 2,500 nautical miles – Poseidon is larger, carries twice the payload and is twice as powerful. Its chief characteristic is that its nose cone separates into multiple, individually-aimed, warheads.

The first sea-firing of Poseidon took place in December 1969 from the experimental ship *Observation Island*, some seven miles off the coast of Cape Kennedy. After this, it was decided to convert the 31 most modern of the US Navy's ballistic missile submarines to carry the new missile, the remaining ten submarines being adapted to carry the A-3 Polaris missile; opportunity being taken of periodical long refits to carry out the necessary conversions, so as to interrupt as little as possible the availability of submarines for patrol. Of the 41 vessels in the Polaris force, about half are on patrol at any given time; the remainder are either at base undergoing a 28-day refit and replenishment between patrols, or in dockyard hands for periodical long refit.

Britain, under the Nassau agreement, has the option of

buying the Poseidon missile from America and converting her Polaris vessels to take it. So far she has not chosen to do so, regarding her A-3 Polaris system as fully effective and likely to remain so for some time. Its British-designed warhead enables the missile to achieve a multiple delivery and there is some confidence that it is capable of meeting every known counter device. But the possibility must be faced that for Britain's purposes, the Polaris weapon may one day prove to be a 'second best' and a change to Poseidon desirable. If so, it will be imperative that the conversion of Britain's four submarines takes place within the confines of each vessel's periodical long refit, when it falls due : only then can the deterrent patrols be maintained. The dockyards are confident that this can be done, barring, of course, unforeseen labour disputes.

Recent American thinking has been in favour of yet another missile which would take the place of the multi-warhead Poseidon. If the SALT talks with the Russians on the limitation of strategic arms were successful it might be possible, some said, to avoid the enormous expense of an entirely new ballistic missile fleet and yet secure a lead by fitting an extended range Poseidon, or Expo, in the 31 vessels of the Lafayette class, with a range of 3,500 to 4,500 miles compared with the 2,500 miles of the Polaris and Poseidon.

The present fleet of Polaris/Poseidon submarines is looked on as the most indestructible of all US strategic weapons, able to come through a nuclear onslaught with the least damage and the greatest capability of mounting a second strike in retaliation. In an all-out effort, it is thought, the Russians could not find and destroy more than one or two ballistic missile vessels on patrol. Nevertheless, plans must be made to accord with any possible major Soviet success in the anti-submarine art.

While an improved missile would give extra protection to a certain degree, a new missile in existing submarines would be far less effective than a new missile in a new submarine. The present Polaris/Poseidon fleet was built in the early and mid 1960's and faces obsolescence in the 1980's : there would be little merit in putting new wine in such old bottles. An entirely new system, comprising new weapons and new submarines would be needed to keep ahead of improved Russian detection and destruction methods.

A new submarine-launched strategic deterrent system has been devised known as the Undersea Long-range Missile system or ULMS, with a range of no less than 6,000 miles. This would mean that Moscow could be threatened by a submarine off the Cape of Good Hope, with similar threats from the South Pacific or Indian Ocean. No anti-submarine forces could compete with such a situation.

The submarines would also be designed to be extremely quiet and to have a high sea-going to harbour ratio. ULMS would at first complement, and finally replace the ten Polaris and 31 Poseidon-armed submarines which will be in service throughout the 1970's. The new submarines would be considerably larger than the existing 7,500 ton vessels, with up to 20 or 24 vertical launching tubes, compared with the present 16. ULMS, an entirely American project, would dramatically increase the United States strategic deterrent. Needless to say, its cost would be enormous : current estimates put the cost of an ULMS force over a ten-year period of $15 to $25 billion according to numbers and equipment carried. Such fantastic sums would seem to put the building of an ULMS submarine force quite out of reach of Britain. But Britain's Polaris force cannot go on for ever and one day the question may well have to be faced of a NATO contribution to some form of Western Allied ULMS Force.

Next in importance is the submerged fire anti-ship weapon, sometimes known as the 'Cruise' missile. Admiral Rickover recently told Congress that 'there is clearly a need for an improved offensive capability to attack surface ships . . . The cruise missile would provide a totally new dimension in submarine offensive capability . . . The very existence of this advanced high performance attack submarine would constitute a threat to both the naval and merchant arms of any maritime force.'

Three main reasons have prompted the US Navy's interest in the anti-ship submarine : the increase in the number and fire power of Russian surface warships; the improvement in Russian anti-submarine capability, and the reduction in US naval forces especially in carriers and carrier-based aircraft. At the present time the offensive capability of US attack submarines is confined to torpedoes, their newest weapon, the Mark 48, having a range of 'more than ten miles'. This compares very unfavourably with the surface-to-surface guided missiles carried in surface ships : the Russian Shaddock missile having a range of about 250 miles. Moreover the vastly slower speed of the torpedo, which has to 'plough' through the water, makes it almost an anachronism in modern warfare. A further advantage of the anti-ship missile for submarines is that the submarine can fire from well outside the reach of any surface ship's counter attack.

The Russians have a clear lead in this field. Over 65 of their submarines are armed with anti-ship cruise missiles, 37 of them nuclear powered and 28 diesel-powered : the 'C' class are each armed with eight long range anti-ship missiles that can be launched while the submarine is submerged.

The proposed American cruise missile submarine, known as the 'Mid-1970 design', primarily intended for an anti-ship role, will also have advanced anti-submarine equipment at least equal to existing attack submarines, making it a dual-purpose vessel, but it is unlikely to be in service before the 1980's.

It is not yet known whether a new submarine will have to be designed around the weapon; much depending on whether the missile can be fired through existing torpedo tubes. It will be fired when submerged, at a target detected by sonar, afterwards breaking surface, to become airborne and finally home on the enemy vessel.

In developing this cruise missile the Americans are in fact picking up the threads of a project they inherited from the Germans at the end of the 1939-45 war. They first launched captured German V1 missiles from shore sites and later an American copy called the Loon. In February 1947 the fleet submarine *Cusk*, which had been given a small water-tight hangar and launching ramp abaft the conning tower, launched a Loon off the coast of California. The *Cusk* was later joined by the submarine *Carbonero* and together they carried out trials with the Loon, achieving ranges of up to 95 miles. A major improvement introduced by the Americans was radio guidance during flight.

The next American advance was a 33 foot missile known as Regulus, resembling a conventional swept-wing jet aircraft flying at nearly the speed of sound for 500 miles, which could be launched with booster rockets from a surfaced submarine. Guided by the launching submarine or by a pre-set electronic brain, and able to carry a nuclear warhead, it was flight tested in 1950 : years ahead of any inter-continental ballistic missile. Moreover, Regulus unlike a ballistic missile could be used against moving targets. Meanwhile, two fleet-type submarines, *Tunny* and *Barbero* were converted into a new class of guided missile submarine, with water-tight hangars abaft the conning tower holding two wing-folding Regulus missiles, and retractable launching ramps. In July 1953 the *Tunny* made the first Regulus missile launching from a submarine.

Encouraged by the success of this programme, a more advanced model, the Regulus II, was designed, 57 foot long with a speed nearly twice that of sound, a range of 1,000 miles and capable of carrying a nuclear warhead.

Like the earlier model it was controlled in flight. Two more submarines, the *Grayback* and *Growler*, while still in the building slips, were converted in 1956 to guided missile craft. This was followed shortly afterwards by the order for a nuclear-powered submarine, the *Halibut*, which became the first submarine in history designed from the keel up as a guided missile vessel. Four more nuclear-powered missile submarines were then authorized. In September 1958 the *Grayback* fired a Regulus II missile off the Californian coast – the first and only submarine launch of this missile. Three months later the Regulus production programme was halted, on account of 'rapidly changing' technology in the missile field; the four guided missile submarines on the drawing board were redesigned as attack submarines and the *Growler*, *Grayback*, *Halibut*, *Tunny* and *Barbero* went to sea with the earlier Regulus I missile.

The *Halibut*, with her large missile hangar incorporated in the bow carried five missiles, *Growler* and *Grayback* carried four each, *Tunny* and *Barbero* two. This guided missile submarine force had a brief life of four years until in 1964 the Navy discarded the whole project, the *Halibut* herself being reclassified as an attack submarine in July 1965. What had happened, of course, was the advent of the Polaris system, technical successor to the Regulus, but with a strategic purpose far more profound.

With the benefit of hindsight, it seems that the US Navy should have carried on the development of the cruise missile side by side with that of Polaris, but Navy planners at the time no doubt thought that the functions of the Regulus missile could be adequately covered by carrier-borne aircraft. The Russians on the other hand, with no carriers of their own, clearly saw the need for this kind of weapon. They started to build their 'E15' class of nuclear-powered submarines, carrying six Shaddock launching tubes, in 1961, followed by the 'E11' class in 1963. Their 'C' class, faster, and carrying a new anti-ship missile which could be fired from underwater, was first reported in 1969.

Britain, in urgent need of an anti-ship missile to provide some 'teeth' for her nuclear-powered fleet submarines, first mentioned in the Defence White Paper for 1968 'studies to increase the effectiveness of submarine launched anti-ship missiles.'

Four years later, in the Navy Debate in the House of Commons, Mr Peter Kirk, Under Secretary for Defence (Navy) admitted that these studies were still continuing. This is a very long time to wait for so important an element in the offensive power of a capital ship – as the fleet submarine is rightly called – leaving aside the long period yet to come of design, development, prototype testing and so on. Whatever the rights and wrongs of this particular programme, it seems to illustrate the difficulties confronting weapon planners in a major new project.

In the first place, when a major ally is working on a similar weapon and appears to be ahead, as the Americans undoubtedly are with their 'mid-1970's design', would it not save time and development costs to buy the completed weapon from them? On the other hand, if our faith in the British design warrants it, allied with considerations of employment for British firms, should we not insist on 'buying British' even if this means a further delay? Or should we compromise, buying the weapon with its main features from America and embodying certain British components, in much the same way as the American McDonnell Douglas Phantom jet fighter was given a Rolls-Royce Spey engine – with the undesirable result that we had to pay more for the aircraft in the end?

These are difficult decisions, which make for delays. There are others inherent in any research and development programme.

After the initial outlay which enables work to begin, a

number of improvements in performance are obtained with little extra money until a point is reached when any significant advance requires the injection of a further very large sum. Should this extra money be forthcoming, the whole process will start again, a number of improvements at modest cost followed by a need for another large sum. It is a zig-zag procession, needing a hard decision as to where development must stop : the courage to say 'thus far and no further.'

An important weapon in the American submarine arsenal is SUBROC (or submarine rocket) designed for use in the ill-fated *Thresher* and the *Permit* and *Sturgeon* classes that followed. A combined rocket and torpedo, it is fired from torpedo tubes like a conventional torpedo, afterwards breaking surface and leaving the water in a ballistic trajectory, finally re-entering the water near the target, when its nuclear warhead is likely to destroy any enemy submarine within the lethal area. At present the weapon is designed for use solely with a nuclear warhead. Taken in conjunction with Soviet claims of nuclear warheads for their submarine torpedoes, this indicates that in the present state of East-West confrontation, tactical nuclear weapons might find a ready use at sea. Studies are in hand, however, for an improved SUBROC anti-submarine missile with a conventional homing head rather than the nuclear warhead. This could be fitted in the new class of high speed submarine, represented by the *Los Angeles* (SSN 688) which will follow the *Thresher – Permit – Sturgeon* series.

The submarine's traditional weapon, the torpedo, achieved its most important development after the American Civil War when the Whitehead torpedo, which ran under its own power, enabled ships to be attacked without incurring suicide. During the 1914-18 war, the submarine's torpedoes were generally of the smaller 18 in type with an effective range of little over half a mile. Usually fired singly, they could sink the larger cruisers and older battleships, but not the more powerful Dreadnoughts. The introduction of the 21 in torpedo gave the submarine a much more lethal weapon.

During the 1939-45 war, the initial successes of the German U-boats in their unrestricted campaign against shipping were gradually offset by the growing efficiency of the Allied convoy escorts. But these ships were small, of shallow draught and able to take rapid avoiding action : a torpedo miss by the submarine usually attracted a dangerous counter attack with depth charges. In 1943, the U-boats started to use an acoustic homing torpedo, attracted by the noise of the escort ship's propellors : whatever avoiding action was taken, the torpedo sought out the noise source. But the innovation was not a surprise to the British Admiralty, which had already devised a simple counter-measure. This was a noise-making device, known as 'Foxer' which was towed astern of the escort and decoyed the torpedo away from the ship. For the rest of the war the convoy escorts had little to fear from the acoustic torpedo. With the advent of the high speed nuclear submarine, the acoustic torpedo lost something of its effectiveness as it can be outdistanced by the submarine. For the same reason, the speed of the nuclear submarine makes it difficult to destroy with depth charges or modern fast-sinking projectiles.

It seems strange that nine years after the *Dreadnought*'s entry into service the main weapon of the nuclear fleet submarine should be a torpedo of pre-war design, the Mark 8. Although viewed with some confidence by submariners as being a thoroughly reliable weapon, which can be brought close enough by the nuclear submarine to prove lethal, in terms of modern weapons it is outdated.

There was a promise of better things when in July 1968 details were announced by the Ministry of Defence of a new 'super' torpedo, effective against all types of submarine, which would shortly come into service as the main armament of the Navy's nuclear-powered fleet submarines and to be carried also by the Polaris vessel. To ensure its publicity, the Press were invited to an Open Day at the secret Admiralty Underwater Weapons Establishment at Portland. This Mark 24 torpedo was described as a 21 in weapon – the same size as its wartime predecessors – wire-guided from the firing submarine, with a wide-angle sonar in its nose which homes on to its target as soon as it comes within detecting range. The guidance system is computerized and the warhead fitted with a proximity fuse which can explode it near the target. An electric engine with the power of a fast sports car was said to give it a high speed and extremely long range, details of which were secret.

But this promising torpedo later ran into production difficulties, showing that it is one thing to produce an almost hand-made prototype with the aid of scientists under laboratory conditions, and quite another matter to start a production line. Some of the complicated electronic work proved to be beyond the capacity of the Navy's torpedo factory at Alexandria, Dumbartonshire and there was a shortage of skilled machinists. The £58 million project, started in 1959, was far too important an undertaking to be abandoned, and it was decided that the work should be taken over by industry, development work being carried out by GEC. Perhaps the best provision for speeding up the programme was the appointment in December 1969 of Sir Rowland Baker, the technical expert behind the Royal Navy's Polaris submarine, as head of a team with wide powers to get production restarted. In the debate on the Navy Estimates in the House of Commons in April 1972, Mr Peter Kirk, Under Secretary of Defence (Navy) said that the Mark 24 was 'now at an advanced stage' and that the weapon would come into service in 1973.

During both World Wars, the gun mounted on the superstructure forward of the conning tower was an important part of a submarine's armament, particularly in the war against merchant shipping, as it provided an alternative to torpedoing without warning. By coming to the surface and engaging the enemy ship with the gun, it was possible to comply with international law by ordering the ship to stop and carrying out visit and search. If contraband was found, the ship could legally be sunk, although the problem of saving the crew remained. An added advantage of the gun attack was that a submarine on a single patrol could destroy a greater number of ships. A U-boat carrying seven torpedoes would be lucky to sink four ships before having to return to base for more torpedoes. With the submarine's greater surface speed it could also overhaul merchantmen sighted over the horizon. For these reasons, U-boats often preferred to attack the slower and smaller ships by gunfire, but it was a method that could be countered by the universal arming of merchant ships and the employment of decoys. With the introduction of radar, submarines on the surface became increasingly vulnerable, and there was a dwindling enthusiasm on their part for attacking with the gun. This led to later designs of submarine dispensing with the gun altogether, thus improving the streamlining of the hull and increasing the submerged speed.

The helicopter now constitutes a growing threat to every type of submarine, particularly when working in pairs – one helicopter finding and 'holding' the submarine, while the other attacks with torpedoes or depth charges. But the helicopter too is vulnerable, especially when hovering to lower its sonar transducer, or 'dipping sonar' into the water. Trials are being carried out with a submarine-launched air missile (SLAM) which consists of a cluster of six 'Blowpipe' anti-tank missiles which can be raised on a mast above the conning tower and controlled from within the submarine by two operators. It seems only a question of time before the submarine will be able to take the offensive against helicopters.

Enough has been said about the choice of weapons in general to show that the introduction of new naval equipment can seldom be swift. Hence the navy with the foresight to undertake preliminary research in peace time is likely to have the advantage in war : experience shows that the lead is likely to be held. Thus the Germans who were in the forefront in the design of offensive submarines and torpedoes at the beginning of the 1939-45 war, were ahead at the end, while the British led in the design of anti-submarine devices throughout.

The choice of what lines of research are to be followed are decided by the naval Chiefs of Staff. In each country the naval staff draw up a list of priorities according to the strategical situation, the country's resources and other general considerations.

Certain major trends will be almost self-evident. Thus Britain with a huge merchant fleet can be sure that any enemy intending to attack her will concentrate on the weapons for attacking these ships : submarines and torpedoes. Britain's answer is to give special attention to research on anti-submarine devices, some of which will be outlined in the next chapter.

CHAPTER EIGHT

The Anti-Submarine Effort

It would be difficult to exaggerate the importance of anti-submarine warfare in any future naval conflict. The threat of the nuclear submarine armed with ballistic missiles, or long range cruise missiles, is such that the central drama in a future naval war will almost inevitably be the struggle between the submarine and its enemies above and below water : the submarine campaigns in both world wars are an indication of this. Had some of the immense air effort devoted to saturation raids on Germany in the last war been applied to the anti-submarine effort, the war might have been significantly shortened. Russia has succeeded Germany in her concentration on submarines, a growing proportion of which are of far higher performance; thereby extending and confirming the cardinal role of the submarine in the balance of power at sea.

A submerged submarine must first be found before it can be attacked and destroyed. The earliest device employed during the 1914-18 war was a light indicator net which showed the position of a submarine caught in it; but this clearly had its limits. Lines of these nets, monitored by drifters, were placed in the Straits of Dover and the entrance to the Irish Sea. Other British counter-measures at that time consisted in laying minefields, establishing a coastal patrol of 500 trawlers and yachts, and the defensive arming of merchant ships.

The patrol vessels had a gun, but some of them also worked an explosive sweep, in the form of a bomb towed underwater astern of the ship, which exploded electrically when passing over a submarine. Another method which had some success was the employment of decoys or 'Q' ships, ostensibly merchant vessels, but manned by the Navy and concealing a heavy gun armament. Most effective of all proved to be the general arming of merchantmen : by September 1916 nearly 2,000 ships had been given a gun. The vast majority of ships so armed escaped when attacked by gunfire.

Meanwhile, British scientists had been studying the properties of the underwater world in order to detect the submarine in its own element. No one can see far through sea water, so light sources were quickly discarded : nor could radio be used, as radio waves are absorbed by water. One possibility was sound, which water conducts well and with little loss over a broad band of frequency.

The hydrophone, an underwater microphone which could pick up the noise of a submarine's machinery or propellors, was the first scientific device capable of detecting a completely submerged submarine. In July 1916 a German U-boat was detected by hydrophones and sunk by depth charges, but the method had a limited value in that all ships in the vicinity had to stop to avoid masking the U-boat's sound. In calm weather the noise of the U-boat and its direction could be plotted by cross bearings.

The detection of submerged submarines by an active sound source, thereby uncovering the completely silent submarine, is said to have been evolved from a suggestion made in 1912, after the sinking of the liner *Titanic* by collision with an iceberg. A British engineer named Richardson suggested that icebergs might be detected by the echo of a pulse of sound waves emitted from an approaching ship. To obtain an indication of the direction of the berg it would be necessary to 'beam' the sound waves like a searchlight, which could only be done with short or supersonic waves. In 1914 the only known practical way of obtaining such waves was from strips of mica made to vibrate under electric stress.

An Allied Committee was thereupon formed to develop anti-submarine techniques, which eventually became known as the Allied Submarine Detection Investigation Committee, giving its name to the ASDIC. It included Dr R. W. Boyle of Canada, Professor W. H. (later Sir William) Bragg, Professor P. Langevin of France and Rutherford. In 1915 Langevin succeeded in producing ultra-sonic waves by applying a technique discovered by Jacques and Pierre Curie. When a quartz crystal is cut in a certain way it will expand or contract when subjected to an electric current : and conversely, when mechanically stretched or compressed it produces an electric charge. An alternating current causes it to vibrate, thereby producing sound waves which can be transmitted through water and reflected back by any obstacle. If the obstacle is sufficiently large, the echo can be received on the quartz which sent out the transmission, creating a weak pulse of alternating current which can be detected by electrical instruments. In the spring of 1918, scientists at the Admiralty Experimental Station at Harwich succeeded with apparatus of this kind in securing supersonic echoes from a submarine at a range of a few hundred yards. The range can be readily determined from the time interval between transmission and echo, based on a speed of sound in water of about 4,000 feet per second. The research work begun at Harwich, was continued by a small group of scientists and engineers at Portland who were able to produce anti-submarine equipment for the Navy by the outbreak of war in 1939 that was far ahead of that possessed by any other nation. The design of the quartz transmitter has been improved by a long series of researches, and much work had to be done to find the best type of dome, or cover for the transmitter, and where it should be placed on the underside of the hull. It was found that the best position for the dome was as far forward as possible on the centre line : if it is off the centre, bubbles of air interfere with the stream past the dome. In submarines it was placed in the bows on top of the superstructure until improved streamlining dictated that it should be inside the submarine's hull.

The active method of acoustic detection represented by the Asdic, or Sonar as it came to be called in the interests of Allied conformity, has a number of advantages. In seeking out the silent submarines, the range as well as the bearing of the target can be accurately known and its movement estimated by measuring the change in frequency, or 'Doppler' effect. A serious disadvantage is that the sound transmissions betray the position of the detecting sonar at a far greater distance than its own zone of detection. This makes the task of the hunter-killer submarine, which relies on concealment and stealth, more difficult.

Another difficulty is that other objects besides submarines return an echo to the sonar pulse, whales, shoals of fish, rocks, tide rips and even water layers of different temperatures tend to confuse the classification of echoes.

Refraction, or the bending of sound waves in water on passing through layers of different temperature, is a special problem : some layers even become impenetrable to sonar. In northern latitudes of the Atlantic, in summer, the top layers of water warm up, decrease their density, and remain on top as a temperature layer. Sonar pulses from ships will not penetrate it, and a submarine can remain undetected below this layer for long periods. When the upper layer of water cools off in winter it sinks and is replaced by warmer, lighter layers from below which in turn cool off and sink. The result is a well mixed-up layer of even temperature extending perhaps 1,000 feet down where sound waves are not deflected and detection is accurate. This was well borne out by the figures for U-boat sinkings in the 1939-45 war, when the number sunk fell off considerably during the summer months.

Other factors can lead to disappointing results such as currents, changes in salinity – particularly near river mouths – and weather conditions. To help predict sonar conditions at various depths, HM ships are supplied with a bathythermograph which is lowered in the water and records the temperature at each depth.

To overcome the difficulty of thermal layers that are impenetrable to sonar, a variable depth sonar has been fitted to some frigates. Largely developed by the Canadian Navy, it enables a sonar set to be lowered to the best operating depth. But the equipment has not been entirely successful; apart from the fact that a long tow astern of the ship can be a positive drawback. No more sets have been ordered from Canada, nor has the Royal Navy developed its own version.

The submarine has a certain built-in advantage in this form of warfare : it is itself a bathythermograph and can use its knowledge of the conditions in its own element to the best advantage, lurking below temperature layers where detection from surface vessels is difficult. It is also immune from the weather disturbances on the surface and is thus a better sonar platform than a surface ship. This greatly adds to the value of the nuclear-powered fleet submarine as an escort for a fast-moving task force.

Active sonar techniques have been improved since the 1939-45 war and a high power transmission is now used, which has increased tenfold the wartime detection ranges. A method of 'bouncing' the sonar transmission off the bottom has been found to increase the range considerably, while the use of 'convergence zones', through which the enemy submarines must pass have enabled the US Navy to claim ranges of 35, 70 and 105 miles, but such results are largely dependent on the right conditions of weather, temperature and depth of water.

Detection of submarines at very great distances by

hydrophones tuned in to very low frequency sounds has been a promising post-war development, and attempts have been made to form off-shore bottom-based ASW barriers to detect the approach of enemy submarines, of which the prime example is the 'Caesar' network, a series of installations at the 100 fathom line off the coast of the United States. Under ideal conditions it has been found possible to listen to a ship's machinery and propellors 300 or more miles away; but good weather conditions are essential and there remains the difficulty of correctly classifying the sounds. This is an area where computers can greatly assist, collating and resolving the information gathered from both sound and oceanographical data.

It would be wrong to discount entirely the anti-submarine function of radar. A surface ship's possession of it tends to 'keep the submarine down' and limits the submarine's own use of radar to supplement its limited perception. Much of the success of the anti-submarine campaign during the later stages of the 1939-45 war was attained by using radar on the many occasions when the submarine had to surface or had to raise its snorting device. Now this possibility has gone, anti-submarine forces have been robbed of their best hope of detecting their target. A brief comparison between sonar and radar, may help to show how the natural properties of water still hamper the scientists who are continually trying to improve the art of underwater detection.

The speed of sonar radiation is slightly under 5,000 feet per second as compared with 1,000 million feet per second for radar. Thus, when a sonar pulse or 'ping' is sent out in a given direction, one minute must elapse before receiving an echo from a target 25 miles away, and a 12 degree beam would take an hour to sweep the full circle of 360 degrees.

The relative movement of searching ship and target has a far greater effect on underwater detection than is the case with radar. A submarine with a submerged speed of

30 knots represents one per cent of the speed of the sonar wave: with radar transmissions 600 knots is only one millionth of the speed of radio waves. Thus the relative movement of submarine and hunting vessel may compress or stretch the received echo by several per cent. This is reflected in the 'Doppler' effect, already mentioned, which has been compared to a gramophone record played at an incorrect and varying speed.

Some novel methods of detection have been tried. Mention has already been made of patrolling aircraft tracing the diesel submarine from its exhaust gases. Attempts have been made to measure the change in magnetic field near the surface of the water when a submarine passes or the temperature of the water when a submarine passes through it. Closely related to this is the theory of the 'thermal scar', which deserves a closer look if only because it is frequently brought forward as a most serious threat to the invulnerability of the Polaris submarine. Founded on a past discovery by American scientists, it is briefly the infra-red detection of a rise in temperature in the wake of a nuclear submarine. The 'thermal scar' thus formed can, it is claimed, be photographed by satellite.

This principle has been known for years but is subject in practice to severe limitations. Submarines have to be very near the surface, otherwise the thermal wake is dissipated : the sea has to be in a state of almost flat calm : cloud interference is also a difficulty. It would be no more wise to consider this method as the death knell of the submarine than it was to give credit to Churchill's wartime dictum 'the Asdic has completely defeated the submarine'. A senior naval officer said recently : 'until someone can devise a new form of wave that can penetrate water as fast and as easily as the radio wave travels through the ether, we shall not get much further.' The United States Navy after eleven years of operating, and many determined hunts of its own believes that its Polaris vessels have gone wholly undetected.

The possibility of detecting the exhaust gases from 'snorting' submarines has already been mentioned. This is done by an instrument that literally 'sniffs' the ionization in the atmosphere : it has been aptly named 'Autolycus', after a gentleman said to have been notorious for picking up 'unconsidered trifles'. At one time, some hopes were pinned on lasers as detecting agents but these are mostly discounted because of the immense energy needed to drive light through water. The heat from a laser with enough power to push the light would boil the water out of its path.

In spite of all efforts, no breakthrough in detection has yet been made. It seems that we must continue to look for gradual improvements to existing techniques, helped by an increasing knowledge of the science of oceanography, which will be discussed in the following chapter.

Once the submarine has been found, there are a number of ways in which it can be attacked. At the outbreak of the last war, the only anti-submarine weapon in service was the depth charge, usually thrown in patterns of five and set to detonate at varying depths between 50 and 500 feet. This had changed little since 1918, apart from an improvement to the firing pistol, preventing the charge from exploding on impact with the water. In the 1939-45 war, British air and ship-borne depth charges accounted for 42.8 per cent of all German U-boats sunk. But the slow rate of sinking of the depth charge, some 10 feet per second, later improved to 16½ feet per second, which enabled the laying ship to evade the explosion of the charge, was often the cause of the submarine escaping damage. A mortar was later introduced, called the Hedgehog, which threw 24 projectiles ahead of the ship with a rate of sinking of 22 feet per second. But the Hedgehog charges had contact fuses, which did not explode unless the U-boat was hit, and towards the end of the war a three-barrelled mortar named Squid was introduced which threw larger charges detonated by time fuses; but this again was an ahead-firing weapon. A further improvement was the Limbo anti-submarine mortar, which could be trained on any bearing. But these mortars, with a range of only a few hundred yards, have not matched recent improvements in detection ranges.

One method of bridging the gap between the range at which a submarine can be detected and the throw of the anti-submarine mortar, is the torpedo-carrying Wasp helicopter, soon to be superseded by the Lynx. Such helicopters are now standard equipment in the larger frigates and can operate also at night.

The larger Sea King helicopter, now in service in the Royal Navy, based on an American Sikorsky design, can both detect and destroy, and is therefore a most important weapon system. With the latest sonar equipment and an operations room, 'almost as large as a frigate's' it can also carry homing torpedoes. Too large to operate from the anti-submarine frigates, they will be carried in Britain's one surviving aircraft carrier, the *Ark Royal*, and in the helicopter carrier cruisers *Blake* and *Tiger*, which have been reconstructed to carry four Sea Kings each, and eventually in the new 'through-deck' cruisers. With a range of over 500 miles these helicopters are proving most effective. The British version, built by Westlands, is fitted with two Rolls-Royce Gnome engines.

The US Navy has developed two weapons which throw a torpedo a considerable distance by means of rockets, Asroc for use by surface ships and Subroc, which is fired from submarines. Both make use of solid-propellant rockets which are jettisoned during flight, with the weapon continuing to its target. Asroc can be used with a depth charge instead of a torpedo, while Subroc has the special advantage of being fired from the submarines torpedo tubes. Nuclear warheads can be used with both weapons, thus strengthening the belief that tactical nuclear weapons might find a more ready use at sea than on land.

The Australian Navy has developed a promising long range anti-submarine weapon system based on a method pioneered by the French. Known as Ikara, it flies an acoustic homing torpedo to the target area attached to a form of radio-controlled expendable drone aircraft, the torpedo entering the sea by parachute. It is in service in the three American-built *Perth* class destroyers and six other Australian destroyer escorts, and will be carried in HMS *Bristol,* the new Type 82 guided missile destroyer.

The influence of aircraft in anti-submarine operations was at its greatest in the latter half of the 1939-45 war when they acted in support of convoys. Fitted with greatly improved radar, they were able to detect submarines on the surface by day or by night and thus foil their attempts to get into a firing position by a fast dash on the surface. But the aircraft's advantage was much reduced by the advent of the 'snorting' submarine towards the end of the war, and almost completely nullified when the nuclear submarine severed all contact with the surface.

The adoption of 'sonobuoys', or small floating sonar sets, dropped by and monitored from aircraft is a method still used by maritime aircraft, and is effective when the U-boat's position is broadly known : they can be either active, sensing out a sonar transmission, or 'ping'; or passive, a purely listening device. But the detecting range of the sonobuoy is limited, its life short, and no very great hopes can be placed on it.

Fixed-wing aircraft are still of great importance in anti-submarine warfare and form a powerful defence for a task force or convoy. They are of particular value as a counter to submarines with cruise missiles, or missile-armed fast patrol boats. They provide the instant response that a surface fleet always requires. If an enemy submarine surfaces 100 miles away to use its cruise missile, the Phantom fighter can be there in five minutes. The US Navy still has enough aircraft carriers to earmark some of them exclusively for anti-submarine duties, but their number has recently been reduced from six to three.

An adaptation of the Comet airliner has provided the long-range maritime-reconnaissance Nimrod, operated by RAF Strike Command. The successors of Coastal Command's Shackletons and twice as fast, they carry a formidable array of detection equipment including sonar, radar, Autolycus ionization detector, magnetic anomoly detector, electronic countermeasures and the invaluable searchlight.

Mines can also be a grave hazard for submarines if the waters in which they are laid are not too deep, and the mine barrier is small enough to be kept under observation, either from the air or by surface craft to prevent the submarine from escaping on the surface. Great ocean mine barrages, such as that laid between the Orkneys and Iceland in the 1939-45 war were remarkably ineffective. Originally planned to be laid between Scotland and Norway a new area had to be found for it after the fall of Norway. It was a costly operation in terms of the conversion of merchant ships to minelaying, and the diversion of much-needed Home Fleet ships for their protection, and resulted in only one U-boat, U703 being sunk, while nine allied merchantmen and one escort ship fell victims to two minefields through navigational errors. The mistake is all the more inexcusable in that a similar mine barrier had been laid in the 1914-18 war between Norway and Scotland with equally disastrous results. It once again prompts the query of why do we not learn from history? On the other hand, mines laid in the entrances to U-boat bases or narrow channels, usually by aircraft, accounted for many successes.

A word must be said about the human factor in anti-submarine warfare. The world's navies have long recognized the need for specialist training in this art and officers and men have been set aside for it and given special qualifications. These were always at hand to advise a commanding officer on the submarine threat, as were the gunnery officers, the torpedo officer the air specialist and so on. But in these days of a continuous all-round threat, from the air, undersea and surface, the carefully drawn lines of specialization in the Navy need to be reviewed. Some officers are now being trained to cope with a wider range of tactical situations, so that in a multi-threat situation, one officer instead of several will advise his captain on the various air, surface and underwater weapons likely to encountered. The speed and complexity of modern warfare require an 'across the board' knowledge rather than deep understanding of a single subject. This concept of a Principal Warfare Officer, pioneered by the Royal Navy, seems likely to commend itself to other modern navies.

CHAPTER NINE

Oceanography and the Submarine

The nuclear submarine has done more than any other single factor in the last decade to stimulate knowledge of the salt water depths that cover 71 per cent of the earth's surface. Man's ability to move, live, and even fight, deeper down than ever before calls for a greater understanding of the sea at all levels. The ocean has an average depth of 12,000 ft, but scientific information is scanty beyond 900 ft – the limit of bathythermograph observations. Submarine operations, until recently confined to a few hundred feet are taking place further down, and if navies are to fight at greater depths they would do well to comprehend the environment right down to the ultimate depth of 36,000 ft.

Oceanography is the study of the ocean and all its boundaries, surface, seabed and coastline. It involves knowledge of the physical, chemical and biological processes which occur there, without which the best use cannot be made of the ocean's resources, nor weapon systems adequately developed. A rough and ready sea lore, inspired by a genuine love of the sea and respect for its power, is no substitute for scientific knowledge. After a lifetime spent at sea, many a seafarer has no very profound knowledge of his element. The best results in oceanography have been obtained through a close liaison between the practical seaman and the civilian scientist. Two hundred years ago, the Royal Navy exemplified this in the fields of navigation and discovery in the persons of Captain James Cook and Sir Joseph Banks; and again 100 years later, in that brilliant start to ocean study – the HMS *Challenger* expedition: the first round-the-world oceanographic expedition ever made. A brief account of this cruise will give a clearer idea of how sailors and scientists can work together.

The *Challenger*, a full-rigged corvette of 2,306 tons, with auxiliary steam power and most of her guns removed to make way for scientific equipment, left Portsmouth on December 21, 1872, under the command of Captain George Nares, passed down Channel in a smooth sea and light headwinds, bound for Lisbon. She soon ran into gales and heavy seas and reached Lisbon on January 3, 1873. The scientific work, under the aegis of the Royal Society, was to be 'of three or four years' duration, during which soundings, thermometric observations, dredging and chemical examination of sea water should be carried on continuously, with a view to the more perfect knowledge of the physical and biological conditions of the great ocean basins, and in order to ascertain their depth, temperature, specific gravity and chemical character . . .' Professor Wyville Thomson, a Scottish naturalist, headed the scientific staff, assisted by John Murray, three other naturalists and a chemist. Biological and chemical laboratories, a photographic dark room and a studio were provided.

Hemp line was still in use for the soundings, as difficulties had been experienced with Lord Kelvin's wire sounding machine, specially designed for the cruise. Soundings and samplings from the depths were carried out over the side, using the main yardarm as a boom, power being provided by a small 18 hp steam engine. Under 1,000 fathoms, the soundings were obtained with the usual lead weight, armed with tallow to secure a bottom sample. For greater depths there were detachable weights of 300 pounds, and a tube for the samples – the forerunner of the modern 'corer'. Depths of 26,850 ft were obtained in the Marianas Trench in the Pacific. Every 200 miles or so, the ship would stop for observations and keep head to wind under her auxiliary steam power: temperatures and bottom samples would then be taken, and specimens of marine life brought to the surface with the trawl. During the three-and-a-half years voyage, 362 of these 'stations' were made, the passage between them being made under sail.

When the *Challenger* returned to England on May 24, 1876, she had completed the longest continuous scientific undertaking of its kind, and one which has never been surpassed in significance. The soundings and bottom samples obtained formed the first complete framework for determinig the basic shapes of the great ocean basins, and are valid to this day.

The scientific reports made by Sir Wyville Thomson and his deputy, Sir John Murray, who succeeded him, took 20 years to prepare, but the 50 quarto volumes which contained them laid the foundations of world oceanography. Considerable differences were found, for instance, in the temperatures and salinities of various layers in the major oceans; currents of some strength and in directions never before suspected, were traced. The discovery for the first time on the ocean floor of deposits such as manganese, iron oxides and phosphorite opened up a new field of marine geology and have been of economic interest ever since. The implications for anti-submarine warfare of the ocean's temperature layers were discussed in an earlier chapter.

The importance of oceanographic knowledge in matters of defence has only gradually become apparent, but for the last 20 years every major naval weapon has required extensive oceanographic data in the development stage. Missiles that are given test firings at sea need precise surveys of the target area: submarines operating submerged for long periods must have full information on their environment: long range sonar requires knowledge of acoustic conditions extending out to many miles.

The importance of oceanography for defence was well put by Mr B. D. Thomas, the American head of Batelle Memorial Institute, in a speech quoted in International Science and Technology for January 1962:

> 'Consider for a moment the hypothesis that the next war will be won not in outer space but on the bottom of the sea. The Russian boast about their missiles and their 100 megaton bombs; they are very quiet about their submarines. They make a lot of noise about their men in space; they are very modest about an effort in oceanography several times as large as ours. What a brilliant military exploit it would be to send us off to the moon, while they seize the ocean. By some logic I have never been able to understand, it has been asserted that the power that controls the moon can conquer the earth. We might add . . . that the power that controls the oceans can control the power that governs the moon.'

A big step forward in the exploration of the ocean floor came with the design and building of the bathyscaphe (derived from two Greek words meaning 'deep Boat'). The Swiss physicist, Auguste Piccard was convinced that the only way to understand the ocean was to take the scientific eye and mind into the deepest part of it, and supplement the scientific information obtained by instruments. His first bathyscaphe in 1948 made an unmanned test dive to 4,500 ft, which proved his theories and excited much interest. A second vessel, built at Toulon, was to have been turned over to the French Navy after its initial tests, but after starting on the design Piccard left for Italy to build his own vessel, the *Trieste*, completed in 1953. In this vessel, Piccard and his son Jacques made a number of successful dives, eventually reaching a depth of 12,110 ft. Meanwhile, the French completed their Piccard-designed bathyscaphe, known as FNRS III, which, in 1954, established a world record dive of 13,287 ft, which held good until the *Trieste* herself invaded the Challenger Deep six years later. But, in the meantime, the US Navy had purchased the *Trieste* 'to examine possible naval uses for this type of craft'.

Before describing the Challenger Deep dive, we should take a closer look at the bathyscaphe itself, and its method of operation. Basically, it consists of two hulls joined together, which counterbalance each other: the upper consisting of a flotation tank; the lower, a heavy pressure-proof steel observation dome. The upper tank, which is of light construction, contains 34,000 gallons of petrol, which, being lighter than water, causes the vessel to float. Water ballast tanks are provided at each end to give the craft negative buoyancy when required. As the craft submerges, the pressure inside the petrol float remains equal to that of the sea water outside, so there is no tendency for the structure to collapse. To return to the surface, steel shot, amounting to 16 tons of disposable ballast, held in containers at each end of the craft, is gradually released, allowing the buoyancy provided by the petrol to assert itself. The observation dome underneath the float is a steel sphere seven feet in diameter, built to withstand pressures of 800 tons per square inch: in it two observers can work and breathe air at normal pressure. A plexiglass window and observation lights enable the occupants to peer into the intense darkness of the ocean depths. Small battery-powered electric motors propel the vessel at about one-and-a-half knots, while a scanning sonar in the bows gives warning of dangerous obstacles ahead. It is a craft that Jules Verne would have revelled in.

The US Navy, impressed with the possibilities of the *Trieste* for oceanographical research, arranged for a number of trial dives in the Mediterranean in the summer of

1957, with Jacques Piccard as pilot, accompanied by a scientist. The tests firmly established the craft's usefulness, and the US Navy decided to buy her on condition that Jacques Piccard accompany her to America to give instruction on her operation; for work with the craft was to form part of the oceanographic programme of the Navy's Electronics Laboratory at San Diego, California. The *Trieste* duly arrived there in December 1958. The following Spring, six dives were made, piloted by Jacques Piccard, after which the craft was refitted in preparation for the world record dive, planned to take place at its deepest known spot, the Challenger Deep Trench. A new observation sphere, capable of withstanding the extreme pressures at that depth, was built by Krupps of Essen to the same outside dimensions as the earlier one, but with thicker walls. After some preliminary trouble with the new sphere's joints, which finally called for the introduction of an entirely new sealing system, the necessary test dives were made, and all was ready for the assault on the ocean's deepest secret.

The dive into the Challenger Deep was successfully made on January 23, 1960, with Jacques Piccard as pilot, accompanied by Lieutenant Don Walsh, the *Trieste*'s Commanding Officer. Very little time could be spent on the bottom, the recorded depth of which was 35,800 ft, but good external lighting enabled them to see a little of their surroundings. What had been achieved was the capability of reaching any depth in the ocean. After this dive Jacques Piccard ended his contract with the US Navy and returned to Switzerland, Lieutenants Walsh and Shumaker becoming the *Trieste*'s pilots, continuing the Electronics Laboratory's programme for the rest of the year. The greater part of 1961 was spent in refitting the vessel, enabling her the following year to take scientists to sea for acoustic, biological and geological investigations. Improvements were also made to much of her equipment, including the propulsion system, the underwater lights and cameras. At this time too there was a change of pilots, Lieutenant-Commander D. L. Keach, USN and Lieutenant G. W. Martin taking turns at diving the craft and keeping watch as Surface Safety Officer. A back-up crew of nine and an Italian engineer from the Castellamare shipyard where she was built, completed the team. In the winter refit that followed, plans were made for a replacement float for the *Trieste*, based on operational experience. The refit also included a thorough overhaul of the observation sphere : a wise precaution, as the craft was about to perform one of its most exacting missions.

When the news came on April 10, 1963 that the nuclear submarine *Thresher* was reported missing, presumed lost, 270 miles west of Boston, Massachusetts, the *Trieste* was one of the many vessels called upon to assist, but she was the only one capable of taking observers down to the seabed at 8,400 ft, in the area of the submarine's last dive. Unfortunately, the *Trieste* was the whole width of the continent away, at her home base in San Diego, California. How could she best be transported quickly to the east coast? Air and rail travel were both considered, but the task was finally given to a dock landing ship, the *Point Defiance*, which sailed through the Panama Canal, with the *Trieste* stored in her capacious well deck. Meanwhile the Trieste's captain, Lieutenant-Commander Donald Keach, flew to Washington to confer with officers directing the search. *Trieste*'s very limited mobility meant that the *Thresher* would first have to be located by other means : afterwards the *Trieste*, with the help of her sonar, which had a range of 400 yards, could close in for visual observation and photography with her four external cameras.

The bathyscaphe arrived in Boston on April 26 and preparations were at once made to get her into action. She was towed out to sea and made a brief test dive some 60 miles off shore. Plans were also made for her future replenishment in the search area, so as to eliminate the long tow to and from harbour. This involved the recharging of batteries and renewal of steel ballast, petrol and air purification equipment. She was then taken back to Boston to await further developments. Meanwhile, the research ship *Conrad* took countless underwater pictures in the area. One promising photograph, which appeared to show debris from the submarine, was found on closer analysis to be depicting a part of the underwater camera's anchor. Further photographs however, showed unmistakable heavy debris, including an air bottle, a broken pipe and some metal plates, and it was announced that the *Trieste* would conduct exploratory dives in the area, where she accordingly arrived under tow on the morning of June 23. Last minute preparations included the positioning of a landing craft to serve as a communications link and a buoy with electronic devices to help the *Trieste* find her way. At this stage, some Russian vessels appeared to be taking great interest in the proceedings, but were warned to keep clear on account of 'submarine survey work'.

On the morning of June 25, the *Trieste* was boarded by Lieutenant-Commander Donald Keach and a civilian scientist, Kenneth Mackenzie, from the Navy Electronics Laboratory at San Diego. They quickly made their way into the confined space of the observation dome and equipment was given a final check. The flooding of the end tanks of the *Trieste*'s float with sea water – usually sufficient to start the vessel – was not enough, due to the choppy sea, and extra ballast had to be thrown on to her deck from the salvage ship *Preserver* nearby. At last, at 10.35 a.m. the *Trieste* began her hour-long descent to the seabed, 8,400 feet below. On this first dive the vessel had some navigational difficulties and went further east than had been planned, but 'hovering' some 40 feet above the bottom, she was able to search some two square miles. The dive planned for the following day was postponed for 24 hours, partly because the electronic buoy was not ready, but also because Lieutenant George Martin, *Trieste*'s second-in-command, had an injured ankle.

The next dive was made on June 26 with Lieutenant Martin as pilot, Mackenzie again serving as observer. On reaching the seabed the *Trieste* settled about two feet into soft silt clay, from which Martin tried to free her by working the craft's three electric motors. The effort of breaking loose took the craft up 480 feet above the bottom, instead of the planned 40 feet, but during the effort to regain control, the *Trieste*'s sonar picked up a large metal object some 60 feet long, at a place where the *Conrad* had also detected metal. Efforts to re-establish this contact failed, and Martin discontinued operations.

The following day, the *Trieste* again submerged, crewed by Keach and Mackenzie. On the bottom they found a yellow rubber shoe cover, or 'bootee', of a kind worn by nuclear submariners when working in the reactor compartment, to prevent radioactive dust from being carried to other parts of the submarine. The bootee, which was folded, clearly bore the marking 'SSN', denoting an attack submarine, and a number beginning with 5 – *Thresher*'s was 503. Quantities of paper and other light debris were also seen, but there were no signs of the large metal object picked up the day before.

On the *Trieste*'s fourth dive on June 29, an additional observer was carried. Lieutenant-Commander Eugene J. Cash, a member of the search operations staff. Two large craters were sighted and photographed, which caused much speculation as to whether large parts of the vessel could have plummeted into the ocean floor. Another theory maintained that the large amount of debris already formed showed that the vessel broke up on the way down, causing a drag that slowed her descent and lessened the impact on the seabed. Captain Andrews pointed out that there was no mound of silt around the edges of the craters, such as one would expect from an object driven with great force into the ocean floor.

Captain Andrews himself joined Keach and Mackenzie on the fifth dive on June 30, but the descent was beset with mechanical failures affecting the gyro compass and one of the three propulsion motors, shortening the dive considerably. The following day the *Trieste* was towed back to Boston for inspection and repairs. Hull welding, the join between dome and float, and all equipment were thoroughly tested. One significant addition was the securing to the craft's forward ballast hopper of a remotely controlled mechanical arm, worked from inside the dome.

It was six weeks before the *Trieste* could return to the search area, meanwhile, oceanographic research ships crossed and recrossed the search area with underwater cameras, sonar and magnetic detectors. One oceanographic tug, the *Allegheny* dropped over 1,000 coloured plastic 'signposts', forming avenues that led to the *Thresher*.

On August 15, ready for sea and under tow, the *Trieste* started for the search area. Bad weather intervened, and it was not until August 23 that the first dive in the new series was made, with Lieutenant-Commander Keach as pilot, and the faithful Mackenzie as observer. Some difficulty was experienced in penetrating a cold water layer at 200 ft, and there were some minor faults in the navigational equipment. After an hour and a half on the bottom, the craft returned to the surface, having at last located one of the plastic markers laid earlier by the *Allegheny*.

There was more bad weather to come and even a serious threat of Hurricane Beulah heading for their area, which caused the *Trieste* to be towed away in the direction of Nova Scotia to escape it. But the hurricane passed harmlessly by, leaving the *Trieste* free to make the second dive of the series. Nothing significant was found on this dive, but the next two, on August 28 and 29, each lasting nearly four hours, were to prove exceedingly important. The fifth and last dive on September 1 was cut short by a failure in one of *Trieste*'s batteries.

Although not announced by the Navy until a week later, it was on August 28, the third dive of the series, that the *Trieste* performed one of the most remarkable feats of her career: the recovery from the seabed at 8,400 feet of a structural part of the *Thresher*. On this occasion, Lieutenant-Commander Keach, the pilot, was accompanied by Lieutenant-Commander A. R. Gilmore, the operations officer on the staff of the search commander, and Commander J. W. Davies of the Navy Electronics Laboratory.

Keach later described his guiding of the *Trieste* through the debris-littered area as being like a visit to 'an automobile junkyard'. Extreme care was needed not to foul the twisted steel wreckage, which could have caused grave danger to the bathyscaphe, or an entanglement, with almost certainly fatal results. The key to the success of the operation was the mechanical arm, fitted to the vessel at the last refit. Keach complained that he had never before succeeded in picking up anything – 'not even a starfish'. It was difficult to decide what to recover: for one thing, the size of underwater objects is not always easy to determine. His first idea was to loop the mechanical arm through a piece of debris, but he decided against this, finally choosing a twisted section of copper pipe, which had certain attachments, lagging insulation, connections, and the like, so that if the arm lost its grip, on the way up, something else might hold.

Very slowly Keach brought the bathyscaphe up to the surface, the ascent taking almost two hours. The pipe, which carried serial numbers, and the designation '594 boat' was identified as part of the air filtration system for the *Thresher*'s galley.

Other large pieces of metal were sighted and photographed, including portions of the hull carrying draft marks, a hatch cover, battery plates and portions of the sonar equipment from the *Thresher*'s forward compartment. The discoveries established that the submarine had at least partially disintegrated, scattering wreckage over a wide area of the seabed; but whether the break-up was caused by some internal disaster or by the tremendous pressure of the depths was not yet known. Further dives by the *Trieste* would not be possible in the North Atlantic winter; moreover the vessel herself needed an extensive refit, including a new float. The Navy Department accordingly announced the ending of the operational aspects of the search.

A Court of Inquiry into the disaster was convened at Portsmouth Naval Shipyard, where the *Thresher* had been built. Testimony was heard from the submarine's former commanding officer, Commander Dean Axene, from two members of the *Thresher*'s crew who had been left ashore on the fateful day, including the ship's electrical officer, from dockyard officals concerned with her refit, and from members of the crew of the rescue ship *Skylark,* who had heard the *Thresher*'s last messages. There were minor differences in the reports of these messages. Lieutenant-Commander Stanley Hecker, the Commanding officer of the *Skylark,* remembered it as: 'Experiencing minor problem. . . . Have positive angle. . . . Attempting to blow', and described the voice from the *Thresher* as 'very relaxed'. All those listening from the rescue ship heard the sounds of air being blown into the submarine's tanks to give her more buoyancy. Lieutenant James Watson, *Skylark*'s First Lieutenant and Navigating Officer, distinctly heard the sounds of a ship breaking up, 'like a compartment collapsing : a muted dull thud'.

Answering widespread fears about the nuclear contamination aspect of the disaster, Admiral Rickover told the court that the reactor used in both submarines and surface vessels were designed to minimize nuclear hazards. 'In the event of a serious accident', he said, 'fuel elements will remain intact and none would be released.' Samples of the ocean bottom taken at the *Thresher*'s known position, he said, showed the seabed to be clear.

Two and a half months after the hearing began, the US Navy announced officially that 'a flooding casualty in the engine room is believed to be the "most probable" cause of the sinking of the nuclear submarine USS *Thresher*. . . . The Navy believes it most likely that a piping system failure had occurred in one of the *Thresher*'s salt water systems, probably in the engine room . . . within moments she had exceeded her collapse depth and totally flooded. She came to rest on the ocean floor, 8,400 feet below the surface.'

Following the Court of Inquiry, certain recommendations were put into effect. Escort vessels concerned with submarine deep-sea trials were ordered to tape-record all transmissions to and from submarines. Although the *Skylark* could not have prevented *Thresher*'s loss or even assisted her, an accurate record of the submarine's last message *might* have provided a clue to the initial 'minor problems'. In the shipbuilding field, ultra-sonic inspection of high pressure piping systems was extended and improvements made in the techniques of silver brazing the pipe joints. For some time Admiral Rickover had been campaigning against poor workmanship in the shipyards. Only six months before, he had said in a speech : 'On more than one occasion I have been in a deeply submerged submarine when a failure occurred in a sea-water system, because a fitting was of the wrong material. But for the prompt action of the crew, the consequences would have been disastrous. In fact I might not be here today.' The Admiral made a particularly serious charge in the matter of welding electrodes :

'Not long ago we discovered a mix-up in the marking and packaging of welding electrodes which also could have had very unfortunate consequences . . . during the next three months, while we were checking this matter in detail, we detected similar incorrect marking and packaging of electrodes in cans from *nearly every major electronic manufacturer in the United States*'. (Admiral Rickover's italics). The improvements then introduced led to a slight increase in the building time for all United States Nuclear submarines.

The search for the *Thresher* certainly focused attention on the science of oceanography. The work done on submerged navigation in the area and the number of soundings carried out, made it the most surveyed plot of deepwater seabed in the world.

Next to the Americans, the French Navy has devoted the most attention to the development of the deep-diving bathyscaphe. They replaced the Piccard-designed FNRS III with the *Archimede* in 1960, which has made six dives to depths greater than 30,000 feet in the Kurile Trench. Bathyscaphe improvements include the use of new, specially tempered steels or titanium for the observation dome, and new buoyancy materials, such as solid plastics, to replace petrol in the float and thus provide a greater measure of safety.

Britain has, as yet, no deep-diving bathyscaphe : instead, she has concentrated on the small submersible, with more immediate practical applications, and capable of dives anywhere on the continental shelf. Best known of these is the Vickers *Pisces*, a highly manoeuvrable three-man vessel, capable of dives to 3,000 feet. Propelled by two small 3 hp battery driven electric motors, her endurance is four hours at two knots or 15 hours at half-a-knot. With two manipulated arms, one of them a torpedo grab, *Pisces* is specially well fitted for recovery, particularly in the case of trial torpedoes that may have gone astray. The arms also enable the craft to carry out such tasks as bottom coring, laying moorings or operating underwater valves. She is usually accompanied by a special support ship, the *Venturer*, which besides being fitted for oceanographic research, with laboratory, dark room, and accommodation for 15 scientists, can raise and lower *Pisces* with powerful over-stern handling gear.

The interest shown by both the submarine and anti-submarine elements of modern navies in the ocean depths has led to a marked increase in oceanographic activity. The whereabouts of peaks and valleys on the ocean bed, hitherto regarded as only of academic interest, now needs careful charting and documentation. At any time, the nuclear submarine may need to weave in and out of these underwater 'forest glades'. The Russians, with their vast underwater fleet, are fully alive to this and their oceanographic vessels are to be found all over the world. Recently, Russian vessels carried out an extensive survey of the Sicilian Narrows in the Mediterranean, a well-known bottleneck for submarines transitting the area.

Marine biology has long had certain direct naval uses, notably the study of organisms that cause the fouling of underwater hulls and loss of ships' speed, and the recognition of noises made by marine animals that can confuse hydrophone listening. Even more important for the future will be marine's biology's part in man's age-old quest for food. Certainly there are almost untapped riches here. While some desirable species of marine life are over-fished almost to the point of extermination, others are hardly used at all. Countries with a great dearth of protein, such as India, make very little use of the fish abounding in the Indian Ocean. This may be partly due to the difficulty of preserving fish in a hot climate without expensive refrigeration, a matter that is largely remedied by modern methods of dehydration and canning. Protein-rich and long-keeping fish flour could provide a necessary supplement to the diet of a quarter of the world's population, at present grievously undernourished. Marine biology and fishery research have to ensure that this harvest is not only gathered, but conserved and renewed. Modern methods of fishing – attracting fish to swim into a funnel and then pumping them into the hold – can all too easily lead to extinction. The right to exploit the oceans of the world implies a responsibility for their conservation. Perhaps this could lead in time to an International Fishery Protection Force, to which the navies of the principal maritime powers could contribute.

But fish is not the only wealth beneath the world's waters. The *Challenger* expedition of 1872 had dredged up manganese nodules, iron oxides and phosphorite, and the finds have been of economic interest ever since. Recently, the American Bureau of Mines dredged from a depth of 12,000 feet nodules that were heavy with manganese, nickel, cobalt and copper, and there is evidence that the nodules form faster than they can be mined.

Submarines have certain advantages as oceanographical research vessels; for instance, they can collect water samples through various hull openings, enabling temperature and salinity to be more efficiently measured. The US Navy in 1960 converted the *Archerfish*, a wartime submarine, for this purpose and the Soviet Navy has done the same with a 'W' class vessel, the *Severyenka*, designed to support fishing activities in northern waters. Her facilities include underwater television and the ability to send out and recover skin divers when submerged.

Oceanography has for many years been a field where international co-operation has been outstanding; and rightly so, for the work needed to assess the marine resources of the world is too great for any one nation. But as the ability to take more and more out of the sea increases, so will international agreements become more and more necessary.

The freedom of the seas has been jealously guarded through the ages; the study and exploitation of the seas for man's advantage needs equal recognition as a field for common effort. In bringing this about, the indivisible ocean bids fair to wash away some of the worst barriers of mankind.

CHAPTER TEN

The Submarine and Western Defence

We have seen how the nuclear submarine has opened up an entirely new era of underwater warfare. With the speed and agility to out-manoeuvre a modern frigate; unlimited endurance; able to maintain its high speed regardless of wind and weather on the surface, and to cross the Arctic under the winter ice, relying solely on its inertial navigation system, it has put itself almost out of combat reach of its enemies. Not only is the nuclear submarine itself the best platform for most anti-submarine equipment, but when hunted by surface vessels, the submarine can always detect its adversary first.

These qualities are remarkable enough in themselves, but when enhanced by possession of the long range ballistic missile, the result is the world's most formidable weapon system, still virtually invulnerable. But the US Navy, which pioneered this revolution in modern warfare is now hard put to it to keep a technological lead over the Soviet Navy, which has been quick to learn the full implications of nuclear submarine warfare.

Dr Joseph Luns, Secretary General of NATO, speaking in Paris in November 1972, on the eve of preparatory talks for the first European Security Conference, said that the situation in naval strength was particularly disturbing. All the Western navies except the American were 'vastly inferior' to the Russians. 'The Russians now have 100 huge nuclear-propelled submarines in service, 60 of them carry inter-continental ballistic missiles, and their large tonnage permits them to cruise round and round the world. They also have 300 conventional submarines, some of them armed with missiles.' This was a grim thought, he added, when one remembered the small number of submarines with which Hitler began the 1939-45 war.

The number of American shipyards building nuclear submarines has decreased from seven to four, Britain's from two to one, while Soviet yards building these vessels have increased from two to four, or possibly five. In a recent book, full of 'gale warnings' on the Soviet Sea Challenge, Rear-Admiral E. M. Eller, USN, credits one Soviet submarine yard, at Sverdovnsk on the White Sea, as having several times the area and facilities of *all* US submarine building yards combined. It is not surprising therefore that Russia now has more nuclear submarines than America, while her force of ballistic submarines with weapons of comparable range to the latest American Polaris vessels is fast catching up. Only brief comfort can be derived from the thought that the greater part of the 400-strong Soviet submarine force is still far from modern; that only a few are equipped with the most advanced missiles, and that thereby the Western Alliance has a breathing space. The situation is altering all the time, and not in the West's favour : the question clearly arises, are the navies now being built by the major naval powers in the West, and said to 'shaping up to the 1980s', best suited to meet the threat?

The United States has by far the largest warship-building potential in the West. But some naval thinkers are challenging the traditional lines along which their new construction is taking place; new ships replacing old ships of similar type, better armed it is true, certainly more costly, but still essentially leading to a Navy much like the US Navy after the Second World War.

The fixed-wing aircraft carrier still holds pride of place. Admiral Elmo R. Zumwalt, the US Navy's Chief of Naval Operations, said recently: 'We control the seas with our aircraft carriers, capable of sinking surface ships, surfaced submarines, shooting down aircraft and, with our F-14 aircraft (the shipborne fighter, successor to the Phantom), shooting down the missiles from any one of those enemy sources.' Few would dispute this as regards the present: even the Royal Navy, which under the Wilson government suffered a far too hasty run-down of carrier strength, is retaining HMS *Ark Royal* until 1978, while France, operating two carriers, is rapidly taking the place hitherto held by Britain. But is the right sort of provision being made for the future, in view of the fact that the Soviet Navy is circumventing the carrier challenge with its emphasis on the submarine? The carrier may perhaps at the moment rule the waves, but it is becoming increasingly powerless to rule the growing threat from beneath the waves. The carrier task force, screened by powerful missile-armed ships, still holds sway in many naval minds, but it is by no means certain that the carriers will have to meet a direct surface or air challenge. This would not be giving Western naval forces command of the seas by default, for a Soviet submarine fleet, nuclear-powered and armed with submerged-launch anti-ship missiles, could dispute command of the seas with existing surface forces. This bid for command of the seas is one that the Soviet Navy seems to be making, deliberately bypassing modes of warfare that would bring them into direct conflict with leading Western navies.

It could be true that, for a time, carrier forces in a conventional war against major Soviet surface vessels, would be victorious, but there would inevitably be occasions at the outset of such a war when the carriers lacked sufficient escorts, or when groups of vessels or single ships found themselves without adequate carrier support. It is difficult to escape the conviction, shared by many in American and British naval circles, that the ascendancy of the submarine as an instrument of naval warfare is an established fact, and that the days of the strike carriers in their limited and diminishing presence, are numbered. Carrying too many eggs in too few baskets, they provide a small number of extremely valuable targets on which the most intense enemy offensive effort could be focused. With satellite surveillance, each carrier is increasingly becoming a marked vessel from the moment of leaving port, and it cannot be long before the sharks – in the form of enemy nuclear submarines – gather round. To defend each with its own nuclear submarine – the only really effective protection it could have – would be a formidable task as well as a dispersal of offensive effort.

Both the American and British Navies have a number of ships originally designed for carrier protection, equipped with weapons rigidly subordinated to that role and with only small guns to use against other ships.

The most successful anti-submarine vessels in the past have been fast destroyers or frigates, equipped with sonar and anti-submarine mortars and often described as 'anti-submarine frigates'. Britain, however, has ceased to build these narrowly specialized vessels and prefers instead a 'general purpose frigate', exemplified by the successful *Leander* class, embodying long-range air warning radar, close-range anti-aircraft missiles and a lightweight helicopter, fitted with dipping sonar and homing torpedoes. New classes of general purpose frigates are succeeding the *Leanders*. These ships are admirable for the manifold tasks that fall to their lot, not least of which is 'showing the flag', but there are those who maintain that since the primary role of such ships is an anti-submarine one, their continued production on a large scale is a misplaced effort, as they are incapable of fulfilling that role against the modern submarine. The conclusion drawn is that more of our resources should be put into the building of nuclear Fleet submarines.

Britain's existing nuclear submarine force of Polaris vessels and Fleet submarines is a source of pride – although the production rate for the latter of one vessel in 15-20 months leaves much to be desired – but the most serious indictment of our naval policy is the failure so far to provide the superb submarines of the *Valiant* and *Swiftsure* classes with anything more than an obsolescent torpedo of World War Two design. With no anti-ship missile of any description, they are thus virtually toothless.

Science is providing yet another field in which the submarine can assert its superiority. It was once a major problem to direct a submarine to a convoy or naval task force. There is now a project whereby a submerged submarine can call up an overflying satellite for information on the whereabouts of ships in a given ocean area. Already photographic satellites are able in clear weather to pinpoint every surface vessel on the ocean, and with improvements in laser technology, even the weather restriction may be overcome. Communications too have advanced : it is now possible to talk to a submerged submarine by radio broadcasts on low frequencies from a shore station, without the submarine having to show a periscope aerial above the surface.

With so many factors pointing to the dominance of the submarine, there would seem to be a good case for building more than just one nuclear submarine for every four surface warships – the present British rate. In fact, for the next two years, the rate will be nearer one to five, owing to a special Government move to assist the shipbuilding industry by hastening the building of frigates and destroyers.

With acute financial stringency affecting all three Services, the building of nuclear submarines is likely to suffer a further setback from the 'through-deck' carrier programme. This controversial vessel, which is virtually a small carrier of up to 20,000 tons – similar in size to the escort carrier of the last war – gets its name from the unobstructed upper deck needed to provide a short take-off run from 'jump jet' aircraft, as well as a landing platform for Sea King helicopters, some earmarked for anti-submarine work, others for airborne early warning. Extensive communication equipment will give the ship command facilities for a naval task force, or combined operations, and Sea Dart missiles for air defence will be carried. The price, originally estimated at £50 million, is not likely in the end to fall far short of £70 million. Three of these ships are planned – to ensure keeping one at sea – and the first is to be ready in 1978, when the *Ark Royal* is withdrawn, the other two replacing the helicopter carriers *Tiger* and *Blake*. The maximum expenditure on these vessels will also coincide with the peak cost of the multi-role combat aircraft – another disturbing factor. What money will then be left for the nuclear Fleet submarine, the capital ship of the future, and the most important of all in terms of command of the seas?

The usual answer to questions of this sort is that Britain must have a 'balanced Fleet', which means a little bit of everything; aircraft carriers, commando ships, assault ships, cruisers, destroyers, frigates, minesweepers, fast patrol boats and the necessary supply ships – besides submarines. But on a restricted budget this too easily results in not enough of anything. This situation is affecting all three Services and has resulted in a recent Government decision to cut down on research – in itself a dangerous economy – and to give priority to those items of equipment that promise the best foreign sales – not necessarily the best for Britain's defence. These restrictions come at a time when the Royal Navy, withdrawing from its traditional world-wide presence, is concentrating on Europe, and prompt the question whether the time has not come for a more intensive effort towards a European Navy, in which each nation from its limited resources, would contribute its own particular expertize, dictated by its geographical position, industrial resources and the role assigned to it in the Western Alliance.

The case for a common European maritime strategy needs to be examined before we can rightly discuss the part the submarine should play in it. Britain's decision to enter the Common Market has drawn attention, as never before, to the pros and cons of a United Europe, with all the economic and political reorganization that this implies. Added to this is the possibility, however remote, of some progressive withdrawal from Europe on the part of the United States. With budgets suffering to a greater or less degree from inflation, the Governments of Western European nations are reluctant to spend money on Defence, or even say much about it in public, thus tending to foster a casual attitude to the subject: a dangerous moment to be gathering round a table for a European Security Conference with Soviet Russia.

Any arrangement between East and West for 'mutual and balanced' force reductions in Europe will suffer from the inherent disadvantage that while Soviet forces need only move a few hundred miles back to their own borders, American troops will be transported to the other side of the Atlantic, 3,000 miles away. It thus becomes imperative for the Western Alliance to give strong naval support to defend the Northern and Southern flanks of NATO, and especially the latter, should there be any move to withdraw units of the 6th Fleet from the Mediterranean.

Much has been said about building a Navy of the 'right shape and size' for the 1980s, but the truth is that measured against the massive Soviet naval threat, no national navy can, of itself, be of the right shape and size, with national defence budgets subject to severe financial restrictions, a concerted European effort would appear to offer the best solution. An agreed European maritime strategy could mean that countries would concentrate on the type of naval operations and warship construction for which they were best suited. Thus Britain and France, who alone among European nations are able to build and operate nuclear submarines, would continue to do so to the exclusion of other, less essential, types; Holland, Italy and Western Germany would concentrate on frigates, Belgium on minesweepers and so on. This would enable more nuclear submarines to be built for the defence of Europe than would otherwise be the case. This unbalancing of national forces would doubtless meet with considerable opposition : in the case of the Royal Navy, for instance, the Naval Staff are hard put to it to find sufficient frigates for the Five Power Commonwealth Force in South-East Asia and for contingencies such as a 'cod war' in Icelandic waters. But viewed in the larger context of a future struggle for command of the seas, it is at least arguable whether the concerted nuclear submarine effort should not be made.

The ideal framework for such co-operation is undoubtedly NATO, assuming that France, in the post-de Gaulle period, gradually returns to the fold, and that the existing organization, under American leadership, is prepared to sustain an Anglo-French nuclear naval alliance for the benefit of Europe. But NATO is still suffering from the French defection, and there are no immediate signs of any change. To get the measure of the problem, it is as well to examine briefly the French objections, which prompted de Gaulle in 1966 to withdraw from NATO and banish its headquarters from French soil. American preponderance in NATO was seen as an affront to the re-awakened power of France, if not to the growing economic and political strength of Western Europe; there was a strong feeling that French armed forces must be under national command in time of war; and – on a more personal note – the American connection was seen by de Gaulle as hampering his own initiatives in building a form of bridge between East and West. Much of these views still remain – in almost crystalline form in the case of the French Defence Minister, M. Michel Debré, they tend to be reinforced by the growth and apparent success of the European Economic Community. On the other hand, France although she has officially withdrawn from NATO, maintains a strong interest in Western defence, keeps up a liaison with many NATO organizations, co-operates in some degree with the major NATO naval exercises, and, above all, is still a member of Western European Union. There seems no reason why, as a long term aim, attempts should not be made at a more closely-knit European Defence Organization, based on Anglo-French naval co-operation – with particular emphasis on submarines. Lord Carrington, Minister for Defence, clearly had this in mind when he addressed the Conservative Party Conference at Blackpool in October 1972: 'I hope that we shall see the French increasingly involved in European Defence, for Europe cannot do without France and France cannot do without Europe. Western Europe has its own nuclear powers in ourselves and France. I foresee one day that the evolution of Europen defence must include some kind of European nuclear force – though not on a comparable style to that of the United States and without, I would hope, any weakening in the partnership with our United States ally. Of course there are plenty of problems there. But if Europe is to have a greater share in its own defence, we shall have to overcome these problems.'

Although, until recently, the official French viewpoint has been sceptical, especially in the case of M. Debré, discussions between the two countries are said to have begun, in spite of French fears that the project would bring about a rift between Bonn and Paris.

The German difficulty in reconciling themselves to an Anglo-French nuclear rapprochement must not be minimized. They fear that anything that encourages the belief in a European nuclear self-sufficiency would only too readily bring about the removal of the American nuclear umbrella; nor do they welcome the notion of France and Britain constituting a 'nuclear élite', with other Europeans providing conventional forces on a sort of feudal basis. These German susceptibilities can only be met by convincing proof that the Anglo-French connection implies no weakening of the American involvement.

Any Anglo-French nuclear submarine co-operation has certain built-in difficulties, one of the most formidable of which is the imparting of American nuclear information. Britain received technical details of weapons and submarine propulsion under the 1958 Agreement, and is precluded from passing them on to another country without the special approval of Congress, which under the McMahon Act – as the US Atomic Energy Act is called – is dominated by the Joint Committee on Atomic Energy. Even President Nixon, who has gone out of his way to restore amicable relations with France, cannot override the Joint Committee, which is always most reluctant to transfer nuclear information to other countries, particularly in the weapons and submarine propulsion fields. Although Britain has since developed her own nuclear submarine propulsion reactor and designed her own nuclear warheads, it would have to be shown that these had not in some way been derived from American information. 'Proving such a negative,' said Ian Smart in his paper for the Institute for Strategic Studies, 'would be as if a man were asked to prove that he had inherited no characteristics from his great-grandfather – and technology has no illegitimate children.' In any case, some items would be of incontrovertibly American origin, such as information about Polaris missiles and associated equipment, and there is a strong possibility that America would prefer to transfer such information directly to France, rather than licence Britain to act as an agent.

As far as an Anglo-French deterrent is concerned, this could serve to divide Europe rather than unite it, were the fear to become widespread, which is particularly strong in Germany, that neither Britain nor France would, in the last resort, use it to cover attacks on other European members of the Western Alliance. An editorial in *The Times*, discussing this very point, concluded that: 'Since the consequences of using nuclear weapons would be suicidal – so far as Britain and France are concerned – it is hardly likely that a Government in either country would use them on anybody else's behalf.'

The British standpoint, which coincides with the American and German one, is that any Anglo-French co-operation in the nuclear deterrent field must be credibly represented as reinforcing deterrence of a Soviet attack on any West European country. But this view is not shared by France, whose unshakable belief is that no country will use nuclear weapons to defend another.

This would seem to show that Anglo-French co-operation in the nuclear strategic weapon field, however much it remains an ultimate ideal, cannot be regarded as an immediate aim, running counter as it does to the present views of the United States, the Federal Republic and other West European countries. But the same can hardly be said of concerted action in the nuclear submarine propulsion field, or in the building and operating of these 'hunter killers', devoid of ballistic weapons. Here Anglo-French co-operation would not offend national susceptibilities in the same way; there would be no question of failing to provide protection to an ally; in fact, a more effective nuclear Fleet submarine force should commend itself as one of the best ways of safeguarding the merchant shipping on which Western Europe as a whole so greatly depends.

The effort and expense to France of producing the five nuclear ballistic submarines of the '*force de dissuasion*' has so far pushed into the background the building of the more generally useful Fleet submarines – just as it slowed down Britain's programme – but there are already signs of a change. The 15-year programme of naval replacement, or '*Plan Bleu*', inaugurated by the French Government in March 1972, includes provision for 20 attack submarines by 1987, of either diesel-electric or nuclear propulsion: the question for the moment is open, but French professional naval pressures for the nuclears are likely to be strong – if only to provide a much-needed backing for their ballistic submarine force.

It is estimated that Britain is some five years ahead of France in the nuclear submarine business, and much advantage would accrue to France from a close liaison in the design, construction and operational fields. Britain, on the other hand, might receive a worthwhile '*quid pro quo*' in the form of help in finding an effective Under Sea Guided Weapon, lack of which is greatly hampering the Fleet submarine's usefulness. France has a special genius in the missile field, recently shown by the fact that the Royal Navy is to begin equipping its surface ships in 1973 with Aerospatiale's Exocet, or 'flying fish' surface-to-surface medium range weapon. Pooling of Anglo-French resources on these two closely associated projects, the nuclear Fleet submarine and the Under Sea Guided Weapon, could prove of immense value to the Western Alliance as a whole.

A move to bring about Anglo-French co-operation in a joint European nuclear-propelled Attack submarine force is not new. It was tried at the 12th Session of the Assembly of Western European Union (of which France remains a member) in 1966, when a report on the subject was tabled by Earl Jellicoe, a former First Lord of the Admiralty and Britain's representative on the Committee of WEU. The terms of reference were carefully chosen, the force was to be 'Joint European', not 'Anglo-French'; but as Britain and France were the only European nuclear powers, it was not difficut to see where the cap would fit.

Although this particular initiative failed, largely perhaps on account of its timing – Europe was not yet ready for such a move – there is so much of positive value in the ideas expressed that it seems reasonable to hope that the scheme will shortly be revived, in some form.

The preamble drew attention in clear terms to Russia's large and increasingly effective fleet of submarines, many of which were nuclear-powered : convincing proof of the need for NATO to maintain up-to-date naval forces, '*especially submarine forces*' (italics mine), to provide for a full range of responses in any military situation. It was undesirable for Europe to leave sole responsibility for the most sophisticated weapons to the United States, and the Assembly considered that only by pooling their efforts could the European countries afford to produce an an adequate nuclear propelled submarine fleet; 'Unless these national programmes can be drawn together quickly in a collective framework, it seems only too possible, given past experience, that each country will go its own separate and expensive way.' Moreover, the Assembly thought, a joint project of this kind would strengthen the political cohesion of the Alliance and offer considerable technological advantages to the countries concerned. The recommendations were brief and to the point :

1. That it invite the member countries to join together in producing *a fleet of nuclear-propelled conventionally-armed submarines, to be jointly developed, constructed and maintained*; (italics mine)
2. That it make a collective approach to the United States Government to arrange for a full and

mutually-beneficial exchange of the relevant technological information;

3. That they set up a Board of Management with responsibility for the joint development and construction programme, for infrastructure, logistics and training, for the administration of any joint contingent that may be established and for assigning the force to appropriate NATO commands;
4. That it amend the modified Brussels Treaty as necessary, to permit the joint research and production programme to be shared with Germany.

The Council agreed with the Assembly on the necessity for maintaining up-to-date naval forces, including submarines, but pointed out that the matter was really one for NATO to decide. With the disarray caused in NATO by France's recent defection, there could hardly have been a more inauspicious moment: no wonder the attempt failed.

The Assembly's proposals can be seen however, as taking much of the heat out of the discussion. Since the deterrent itself was not in question, America need only be approached for information regarding the submarines themselves and their propulsion systems, which she could either impart to France directly, as with Britain in 1958, with Britain and France co-operating later by separate agreement, or via Britain as an intermediary. Joined with the evident good intentions towards Germany, contained in the final recommendation, the proposals were well-calculated to further the cause of European unity.

Enough has been said of the problems to show that there is no quick or easy solution to Anglo-French co-operation. Perhaps there will be a revival in some form of the 1966 attempt. Meanwhile the cautious men in the Kremlin must inevitably view the growing nuclear submarine strength of France as a force to be reckoned with on the Western side. They have already given notice that for the purposes of the Strategic Arms Limitation Talks between the United States and the Soviet Union, they regard British and French ballistic missile submarines as part of the allowable United States total – an unsolicited tribute to French defence thinking.

A sign that the submarine is becoming of more consequence in NATO planning is shown by a recent decision to include two submarines, one of them a nuclear, in the Standing Naval Force, Atlantic, which exists to demonstrate in an emergency the collective will of NATO, and usually consists of destroyers and frigates. This welcome addition came during the period as Commodore of the Force of a former captain of Britain's nuclear submarine *Dreadnought*.

Never in the history of the world has sea power been more important than it is today; for the Western Alliance, the main bastion of Western Civilization, is in essence a maritime alliance, and one that is being challenged before our eyes by a nation whose children seem wiser in their generation than the children of the West. But the exercise of sea power is rapidly passing over to the submarine. In two world wars the submarine nearly proved the decisive factor, but in the end the hunters won. Now the odds are strongly on the hunted and against the hunter. The dominant reality of naval warfare today and for the foreseeable future is the fighting superiority of the submarine.

Index